Lecture Notes in Computer Science 14293

Founding Editors

Gerhard Goos
Juris Hartmanis

Editorial Board Members

The series Lecture Notes in Computer Science (LNCS), including its subseries Lecture Notes in Artificial Intelligence (LNAI) and Lecture Notes in Bioinformatics (LNBI), has established itself as a medium for the publication of new developments in computer science and information technology research, teaching, and education.

LNCS enjoys close cooperation with the computer science R & D community, the series counts many renowned academics among its volume editors and paper authors, and collaborates with prestigious societies. Its mission is to serve this international community by providing an invaluable service, mainly focused on the publication of conference and workshop proceedings and postproceedings. LNCS commenced publication in 1973.

Lisa Koch · M. Jorge Cardoso · Enzo Ferrante ·
Konstantinos Kamnitsas · Mobarakol Islam ·
Meirui Jiang · Nicola Rieke ·
Sotirios A. Tsaftaris · Dong Yang

Editors

Domain Adaptation and Representation Transfer

5th MICCAI Workshop, DART 2023
Held in Conjunction with MICCAI 2023
Vancouver, BC, Canada, October 12, 2023
Proceedings

Springer

Editors
Lisa Koch (iD)
University of Tübingen
Tübingen, Germany

Enzo Ferrante (iD)
CONICET/Universidad Nacional del Litoral
Santa Fe, Argentina

Mobarakol Islam
Imperial College London
London, UK

Nicola Rieke (iD)
Nvidia GmbH
Munich, Germany

Dong Yang
Nvidia (United States)
Santa Clara, CA, USA

M. Jorge Cardoso (iD)
King's College London
London, UK

Konstantinos Kamnitsas (iD)
University of Oxford
Oxford, UK

Meirui Jiang
Chinese University of Hong Kong
Hong Kong, Hong Kong

Sotirios A. Tsaftaris
University of Edinburgh
Edinburgh, UK

ISSN 0302-9743   ISSN 1611-3349 (electronic)
Lecture Notes in Computer Science
ISBN 978-3-031-45856-9   ISBN 978-3-031-45857-6 (eBook)
https://doi.org/10.1007/978-3-031-45857-6

This Springer imprint is published by the registered company Springer Nature Switzerland AG
The registered company address is: Gewerbestrasse 11, 6330 Cham, Switzerland

Paper in this product is recyclable.

Preface DART 2023

Recent breakthroughs in advanced machine learning and deep learning have revolutionized computer vision and medical imaging, enabling unparalleled accuracy in tasks such as image segmentation, object recognition, disease detection, and image registration. Although these developments have greatly benefited the MICCAI community, many models suffer from limited adaptability when faced with novel scenarios or heterogeneous input data. To overcome this restriction, researchers have explored techniques such as transfer learning, representation learning, and domain adaptation, allowing for improved model training, effective domain adaptation, and the application of knowledge learned from one domain to tackle challenges in other domains. By expanding the versatility and robustness of these cutting-edge methods, researchers hope to increase their clinical utility and broaden their impact across various medical imaging applications.

The 5th MICCAI Workshop on Domain Adaptation and Representation Transfer (DART 2023), which happened in conjunction with MICCAI 2023 in Vancouver, Canada, aimed at creating a discussion forum to compare, evaluate, and discuss methodological advancements and ideas that can improve the applicability of machine learning (ML)/deep learning (DL) approaches to clinical settings by making them robust and consistent across different domains.

During the fifth edition of DART, 16 full papers were accepted out of 32 submissions that underwent a rigorous double-blind peer review process. At least three expert reviewers evaluated each paper, ensuring that only high-quality works were accepted. To maintain objectivity and prevent any biases, reviewer assignments were automated and took into account potential conflicts of interest and recent collaboration history between reviewers and authors. The final decision on acceptance or rejection was made by the area chairs based on the reviews received, and these decisions were always fair and unappealable.

Additionally, the workshop organization committee granted the Best Paper Award to the best submission presented at DART 2023. The Best Paper Award was assigned as a result of a secret voting procedure where each member of the committee indicated two papers worthy of consideration for the award. The paper collecting the majority of votes was then chosen by the committee.

We believe that the paper selection process implemented during DART 2023, as well as the quality of the submissions, resulted in scientifically validated and interesting contributions for the MICCAI community and, in particular, for researchers working on domain adaptation and representation transfer.

We would therefore like to thank the authors for their contributions and the reviewers for their dedication and professionalism in delivering expert opinions about the submissions.

August 2023

Lisa Koch
(Lead Editor)

M. Jorge Cardoso
Enzo Ferrante
Mobarakol Islam
Meirui Jiang
Konstantinos Kamnitsas
Nicola Rieke
Sotirios A. Tsaftaris
Dong Yang

Organization

Editors and Program Chairs

Koch, Lisa (Lead Editor)   University of Tübingen, Germany
Cardoso, Jorge   King's College London, UK
Ferrante, Enzo   CONICET / Universidad Nacional del Litoral, Argentina
Islam, Mobarakol   Imperial College London, UK
Jiang, Meirui   The Chinese University of Hong Kong, China
Kamnitsas, Konstantinos   University of Oxford, UK
Rieke, Nicola   Nvidia GmbH, Munich, Germany
Tsaftaris, Sotirios   University of Edinburgh, UK
Yang, Dong   Nvidia Corporation, Santa Clara, USA

Program Committee

Atapour-Abarghouei, Amir   Durham University, UK
Bai, Wenjia   Imperial College London, UK
Banerjee, Abhirup   University of Oxford, UK
Bdair, Tariq   Technical University of Munich, Germany
Bredell, Gustav   ETH Zurich, Switzerland
Chaitanya, Krishna   Janssen Pharmaceuticals R&D, Johnson & Johnson, Switzerland
Chen, Long   University College London, UK
Dinsdale, Nicola   University of Oxford, UK
Emre, Taha   Medical University of Vienna, Austria
Erdil, Ertunc   ETH Zurich, Switzerland
Feng, Ruibin   Stanford University, USA
García-Ramírez, Jesús   UNAM, Mexico
Gonzalez, Camila   Technical University Darmstadt, Germany
Guan, Hao   University of North Carolina at Chapel Hill, USA
Haghighi, Fatemeh   Arizona State University, USA
Hosseinzadeh Taher, Mohammad Reza   Arizona State University, USA
Hu, Yang   University of Oxford, UK
Ilanchezian, Indu   University of Tübingen, Germany
Jiménez-Sánchez, Amelia   IT University of Copenhagen, Denmark

Karani, Neerav   MIT, USA
Laves, Max-Heinrich   Philips Research, Germany
Li, Jinpeng   The Chinese University of Hong Kong, China
Li, Zeju   Imperial College London, UK
Li, Kang   The Chinese University of Hong Kong, UK
Liu, Xinyu   City University of Hong Kong, China
Ma, Wenao   The Chinese University of Hong Kong, China
Manakov, Ilja   ImFusion, Germany
Mansilla, Lucas   CONICET / Universidad Nacional del Litoral,
                                    Argentina
Menten, Martin   Technische Universität München, Germany
Minh Ho Nguyen, Duy   German Research Centre for Artificial
                                    Intelligence, Universität Stuttgart, Germany
Morshuis, Jan Nikolas   University of Tübingen, Germany
Namburete, Ana Ineyda   University of Oxford, UK
Ouyang, Cheng   Imperial College London, UK
Paetzold, Johannes   TUM, Germany
Paschali, Magdalini   Stanford University, USA
Saha, Pramit   University of Oxford, UK
Seenivasan, Lalithkumar   National University of Singapore, Singapore
Seibold, Constantin   Karlsruhe Institute of Technology, Germany
Seth, Pratinav   Manipal Institute of Technology, India
Sheikh, Rasha   University of Bonn, Germany
Sudre, Carole   University College London, UK
Thermos, Spyridon   University of Edinburgh, UK
VS, Vibashan   Johns Hopkins University, USA
Vuong, Trinh Thi Le   Korea University, Korea
Wang, Rongguang   University of Pennsylvania, USA
Xia, Yong   Northwestern Polytechnical University
Xia, Tian   Imperial College London, UK
Zhang, Yue   Siemens Healthineers, USA
Zhou, Kang   The Chinese University of Hong Kong, China
Zimmerer, David   German Cancer Research Center (DKFZ),
                                    Germany

Contents

Domain Adaptation of MRI Scanners as an Alternative to MRI Harmonization

Rafsanjany Kushol[1]([✉]), Richard Frayne[2], Simon J. Graham[3],
Alan H. Wilman[4], Sanjay Kalra[1,5], and Yee-Hong Yang[1]

[1] Department of Computing Science, University of Alberta, Edmonton, Canada
kushol@ualberta.ca
[2] Department of Radiology, University of Calgary, Calgary, Canada
[3] Sunnybrook Health Sciences Centre, University of Toronto, Toronto, Canada
[4] Department of Radiology and Diagnostic Imaging, University of Alberta,
Edmonton, Canada
[5] Department of Medicine, University of Alberta, Edmonton, Canada

Abstract. Combining large multi-center datasets can enhance statistical power, particularly in the field of neurology, where data can be scarce. However, applying a deep learning model trained on existing neuroimaging data often leads to inconsistent results when tested on new data due to domain shift caused by differences between the training (source domain) and testing (target domain) data. Existing literature offers several solutions based on domain adaptation (DA) techniques, which ignore complex practical scenarios where heterogeneity may exist in the source or target domain. This study proposes a new perspective in solving the domain shift issue for MRI data by identifying and addressing the dominant factor causing heterogeneity in the dataset. We design an unsupervised DA method leveraging the maximum mean discrepancy and correlation alignment loss in order to align domain-invariant features. Instead of regarding the entire dataset as a source or target domain, the dataset is processed based on the dominant factor of data variations, which is the scanner manufacturer. Afterwards, the target domain's feature space is aligned pairwise with respect to each source domain's feature map. Experimental results demonstrate significant performance gain for multiple inter- and intra-study neurodegenerative disease classification tasks. Source code available at (https://github.com/rkushol/DAMS).

Keywords: Dataset heterogeneity · Disease classification · Domain adaptation · Domain shift · MRI

1 Introduction

Techniques in neuroscience research are required to be robust, efficient and reliable. They must be sensitive to biological factors but resistant to non-biological

Supplementary Information The online version contains supplementary material available at https://doi.org/10.1007/978-3-031-45857-6_1.

L. Koch et al. (Eds.): DART 2023, LNCS 14293, pp. 1–11, 2024.
https://doi.org/10.1007/978-3-031-45857-6_1

sources, which include changes due to variations in instruments such as Magnetic Resonance Imaging (MRI) scanners. MRI is a classical diagnostic tool in clinical practice. However, the acquisition process of MRI scans can affect the appearance of healthy and abnormal tissue in MR images, potentially impacting the performance of deep learning (DL) methods. Convolutional neural networks (CNNs) have been outstanding in several medical image analysis tasks, though they can be sensitive to deviations in image acquisition protocols. In other words, a CNN trained on samples with a particular MRI protocol is usually less effective on images from a different protocol because of domain shift [21]. This can limit the applicability of DL models in clinical research settings, where it is common to encounter new sites within a study or to merge data from multiple studies.

Recent years have seen a significant boost in demand for integrating multiple neuroimaging studies, as it offers advantages such as increased sample size, population diversity, and statistical power. However, in a clinical research setting, several challenges arise when incorporating multi-center datasets, such as lacking labelled annotations in the target domain as well as the use of different MRI scanners in the source dataset. Moreover, the target domain's data may be heterogeneous too, meaning that it may consist of different scanners causing domain shift within the target domain. To our knowledge, no existing study has concurrently addressed these practical aspects of multi-center MRI data.

Although many research studies have demonstrated exemplary performance on certain domains and MRI protocols, it is often not assessed whether these methods can be generalized to target data with different imaging distributions [27]. To ensure that the trained models can be used effectively in real-world clinical practice, it is essential to overcome the aforementioned challenges. Towards this, some studies [3,25,28] have used supervised domain adaptation (SDA) techniques that require labels from the target domain. A group of works [8,30] have trained their models on source domains and fine-tuned the pre-trained models with partially labelled data from the target domain in a semi-supervised fashion. However, some solutions [18,19] have only considered single-source DA, while many [1,26] have not considered the heterogeneity in the target domain.

In a multi-center MRI dataset, domain shift refers to the differences in scanners and imaging protocols across different sites. Some examples of domain shift parameters include imaging protocol (flip angle, acquisition orientation, slice thickness) and scanner (manufacturer, model, field strength). Therefore, MR images may differ qualitatively from center to center and study to study. Dealing with all of these parameters can be computationally complex and may not be necessary. In several MRI studies, we observe that the dominant factor responsible for data deviation is the scanner manufacturer, as shown in Fig. 1 using the t-SNE [16] method. Indeed, our results confirm a prior analysis [24], which considered several underlying factors of site effects but reported the scanner manufacturer as the most significant parameter causing site effects.

The present study introduces a new perspective to address the domain shift issue in multi-site MRI data using unsupervised domain adaptation (UDA). The prime task in UDA is to learn with labelled source samples, and the objective for the model is to function satisfactorily in a separate target domain without labels.

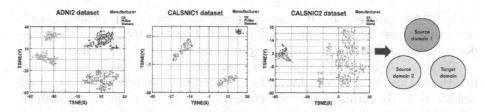

Fig. 1. Graphs show the distribution of MRI data used in our study from the ADNI [11] and CALSNIC [12] datasets generated by the features of MRQy [22] using t-SNE. Three different colours indicate three different MRI scanner manufacturers' data which are separable from each other. The rightmost panel shows that among three manufacturers, two can be regarded as source domains and the other as the target domain. More findings with different datasets are given in Supp. Figure S1.

By classifying different scanner manufacturers as distinct domains, the proposed framework can learn better domain-invariant representation and enhance cross-domain classification accuracy.

Our novel contributions are:

1. Demonstrating the necessity of appropriate domain selection for adaptation instead of considering an entire study/dataset as the source or target domain.
2. Combining maximum mean discrepancy (MMD) [31] and correlation alignment (CORAL) [23] loss to extract pairwise domain-specific invariant features.
3. Conducting comprehensive MRI experiments to classify individuals with amyotrophic lateral sclerosis (ALS) or Alzheimer's disease (AD) in relation to healthy controls (HC) to evaluate the efficacy of the proposed method.

2 Related Works

Recent years have seen some notable works using supervised, semi-supervised or unsupervised DA techniques in either segmentation [1,8,18,19] or classification [25,26,28,30] task. Ghafoorian *et al.* [8] utilized transfer learning and reported that without fine-tuning with the target domain, the pre-trained model reasonably failed for brain white matter hyperintensity (WMH) segmentation. A mixup strategy-based UDA [19] for knee tissue segmentation revealed that the model trained from scratch with fewer samples lacked generalization and performed worse on different domains. Another UDA [18] used a paired consistency loss to control the adaptation for WMH segmentation. Neither of these studies considered multiple source domains nor heterogeneous target domains.

Wachinger *et al.* [25] designed an SDA-based classification model for AD detection by regularizing the multinomial regression employing $l1/l2$ norm. Another SDA framework [28] ignores scanner bias while focusing on pathology-related features for HC vs. multiple sclerosis (MS) patients classification. With pre-training and fine-tuning phases, Zeng *et al.* [30] utilized federated learning for schizophrenia and major depressive disorder classification tasks. Another SDA study [3] generated scanner invariant representation for the age prediction task. Note that these methods require full or partial ground truth from the target domain, which may be time-consuming and costly in a real-life scenario.

Wang *et al.* [26] developed a pre-trained classifier with the source data and fine-tuned this model to new data, which showed improvement in the AD classification task. By learning a common embedding with source and target samples, the UDA framework Seg-JDOT [1] was developed for MS lesions segmentation. These studies ignored the heterogeneity in the target domain dataset. More specifically, data variations may exist in the target domain due to using different MRI scanners; hence this heterogeneity should be handled appropriately.

Many efforts have been made to address the negative impact of scanner bias using MRI harmonization, which aims to mitigate site effects while retaining the statistical power to detect biological factors in images. Image translation-based methods like CycleGAN or neural style transfer render harmonized images to address the issue of domain shift [15]. Statistics-based techniques such as Com-Bat [5] harmonize biomarkers extracted ROIs to alleviate scanner bias. However, a recent study [7] discloses that existing image translation or statistical approaches failed to harmonize cortical thickness from multi-scanner MRI data properly. Moreover, modifying pixel intensities before training may not be ideal for medical applications as they might remove meaningful pixel-level details needed for different tasks such as anomaly detection. Unlike existing harmonization methods, which apply transformation to images to reduce scanner bias, the proposed method adapts datasets from various scanners without changing the images.

3   Methods

Assume that we have N source domains with labeled samples $\{\mathcal{X}_s^j, \mathcal{Y}_s^j\}_{j=1}^N$, where $\mathcal{X}_s^j$ denotes data from the jth source domain and $\mathcal{Y}_s^j$ are the corresponding class labels. Additionally, we have target domain $\mathcal{X}_t$, with unlabeled $\mathcal{Y}_t$. UDA aims to learn a model that can generalize well to the target domain while minimizing the domain shift between the source and target domains. Specifically, given the source and target domains, the objective is to learn a domain-invariant feature representation $\mathcal{F}$ that can capture the underlying data distributions across different domains. To achieve this, the discrepancy between the source and target feature distributions must be minimized while maintaining the discriminative information necessary for downstream tasks, such as classification.

3.1   Proposed Method

The overall workflow of the proposed approach is shown in Fig. 2. In stage 1, the proposed framework classifies the domains based on scanner manufacturers (GE, Siemens, Philips). Scanner information is readily available in a standard MRI dataset [26]; however, if this is not available, t-SNE features can be generated using open-source MRQy [22], which can be clustered using K-means clustering

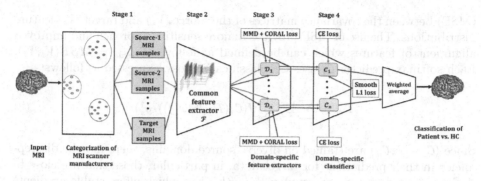

Fig. 2. The overall workflow of the proposed stages. Best viewed in colour.

to categorize the data based on scanner manufacturers. In stage 2, the common latent feature representation $\mathcal{F}$ is determined from the original feature space of all available domains. Although some methods try to extract domain-invariant features in this shared space, in practice, extracting domain-invariant features across multiple domains leads to a higher degree of discrepancy. Motivated by [31], $\mathcal{F}$ is extended to multiple feature spaces (stage 3), aligning the target domain with available source domains by learning multiple domain-invariant representations $(\mathcal{D}_1, \cdots, \mathcal{D}_n)$ by minimizing the MMD and modified CORAL loss. Using these pairwise domain-invariant feature maps, an equal number of domain-specific softmax classifiers $(\mathcal{C}_1, \cdots, \mathcal{C}_n)$ are trained in stage 4, which exploit the Cross-Entropy (CE) loss on the labels of source domains. Finally, to minimize the dissimilarity in each $(\mathcal{C}_1, \cdots, \mathcal{C}_n)$, the $Smooth_{l1}$ loss function [9] is employed which is less sensitive to outliers. The final target decision (Patient vs. HC) is generated from the weighted average of the outputs of $(\mathcal{C}_1, \cdots, \mathcal{C}_n)$ following [20]. The advantage of having multiple classifiers is that if there are fewer samples from a particular manufacturer, then the data from other manufacturers can be used to achieve better performance.

3.2 Loss Functions

To align the feature space of the source and target domains, we leverage the joint contribution from the MMD and deep CORAL loss functions. MMD measures the distance between the empirical mean embeddings of the source and target domains in a reproducing kernel Hilbert space, and the details of $\mathcal{MMD}()$ can be found in [31]. Each feature extractor $(\mathcal{D}_1, \cdots, \mathcal{D}_n)$ learns a domain-invariant map for each pair of source and target domains by minimizing $\mathcal{L}_{mmd}$ as follows:

$$\mathcal{L}_{mmd} = \frac{1}{N} \sum_{j=1}^{N} \mathcal{MMD}(\mathcal{D}_j(\mathcal{F}(X_s^j)), \mathcal{D}_j(\mathcal{F}(X_t))) \tag{1}$$

Deep CORAL [23] aims to minimize the difference between the source and target covariance matrices (second-order statistics) in a d-dimensional feature space. We replace the Frobenius norm and normalization term with the mean squared error

(MSE) between the covariance matrices of the source (V_s) and target (V_t) feature distributions. The use of MSE provides a more sensitive approach and improved alignment of features which can be defined as follows: $\mathcal{L}_{coral} = MSE(V_s, V_t)$. Each softmax predictor $\mathcal{C}_j$ uses CE classification loss, expressed as follows:

$$\mathcal{L}_{ce} = \sum_{j=1}^{N} \mathbb{E}_{x \sim X_s^j} J(\mathcal{C}_j(\mathcal{D}_j(\mathcal{F}(x_s^j))), y_s^j) \tag{2}$$

Since $(\mathcal{C}_1, \cdots, \mathcal{C}_n)$ are trained on diverse source domains, there may be discrepancies in their predictions for target data, in particular, those that are close to decision boundaries. So we employ $Smooth_{l1}$ loss which offers stable gradients for larger values and fewer oscillations during updates to yield a similar classification from each $\mathcal{C}_j$ for the same target sample. Finally, the total loss is noted as: $\mathcal{L}_{total} = \mathcal{L}_{ce} + \lambda(\mathcal{L}_{mmd} + \mathcal{L}_{coral} + \mathcal{L}_{Smooth_{l1}})$, where λ is the adaptation factor.

3.3   Implementation Details

The proposed framework employs 32 coronal slices from the central plane of 3D MRI. The final class prediction is performed with the majority voting of these coronal images, similar to [14] [13]. This slice range can better capture significant brain regions related to AD/ALS, including hippocampus, motor cortex, and corticospinal tract. We use ResNet-50 [10] as the backbone, which is pretrained on ImageNet [2]. When both training and testing occurred on a specific domain, we split the data as train:validation:test with a 6:2:2 ratio, and mean classification accuracy is reported from five repeated experiments with randomly shuffled data. Avenues of data leakage are avoided following [29]. We use minibatch SGD optimizer with a momentum of 0.9 and a batch size of 32. Using $\eta_p = \frac{\eta_0}{(1+\alpha p)^\beta}$, where $\eta_0 = 0.01$, $\alpha = 10$ and $\beta = 0.75$, we adapt the learning rate during SGD. Instead of fixing the adaptation factor λ, we modify it from 0 to 1 with iterative scheduling: $\lambda_p = \frac{2}{exp(-\theta p)} - 1$, where $\theta = 10$, to reduce early training phases noisy activations following [6]. We use NVIDIA RTX A6000 GPU, which took around 8 h to train on the largest dataset (ADNI1).

4   Experiments

4.1   Datasets

Publicly available longitudinal datasets from the Alzheimer's Disease Neuroimaging Initiative (ADNI) [11], Australian Imaging, Biomarker and Lifestyle (AIBL) [4] and Minimal Interval Resonance Imaging in Alzheimer's Disease (MIRIAD) [17] are used for classifying AD patients. The Canadian ALS Neuroimaging Consortium (CALSNIC) [12] multi-center dataset is used for classifying ALS. For ADNI and CALSNIC, two independent versions are used, ADNI1/ADNI2 and CALSNIC1/CALSNIC2, respectively. We use T1-weighted structural MR images for the above datasets, which are skull-stripped and registered to MNI-152 standard space using the FreeSurfer and FSL software, respectively. After preprocessing, the resulting image dimension is $182 \times 218 \times 182$, and the voxel dimension is converted to 1 mm isotropic resolution. Each dataset's demographics and scanning details are given in Table 1 and Supp. Table S1, respectively.

Table 1. Demographic details of the ADNI, AIBL, MIRIAD and CALSNIC datasets

Dataset (#total)	Group	GE Sex (M/F)	GE Age (Mean±Std)	Siemens Sex (M/F)	Siemens Age (Mean±Std)	Philips Sex (M/F)	Philips Age (Mean±Std)
ADNI1	AD	80/80	75.5 ± 7.7	80/80	75.0 ± 7.2	60/49	75.7 ± 7.0
(925)	HC	80/80	75.1 ± 5.7	80/80	75.9 ± 5.9	109/67	75.4 ± 5.2
ADNI2	AD	61/41	75.0 ± 8.5	90/57	75.1 ± 7.8	48/58	74.5 ± 7.3
(852)	HC	64/90	74.3 ± 5.9	92/88	74.0 ± 6.4	69/94	75.6 ± 6.4
AIBL	AD	–	–	28/45	73.6 ± 8.0	–	–
(288)	HC	–	–	99/116	72.9 ± 6.6	- -	–
MIRIAD	AD	19/27	69.4 ± 7.1	–	–	–	–
(69)	HC	12/11	69.7 ± 7.2	–	–	–	–
CALSNIC1	ALS	21/25	57.0 ± 11.4	43/28	59.6 ± 10.8	17/1	58.1 ± 9.0
(281)	HC	23/33	50.5 ± 11.9	38/28	57.2 ± 8.1	6/18	53.1 ± 8.4
CALSNIC2	ALS	14/4	54.0 ± 11.8	124/65	60.1 ± 10.2	29/20	62.4 ± 8.2
(546)	HC	18/13	60.1 ± 8.8	120/101	54.9 ± 10.5	12/25	61.7 ± 10.8

4.2   Intra-study Validation

While merging a new site into a study, the most challenging scenario occurs when the scanner manufacturer of the new site differs from those currently involved. To assess the potential impact of such extreme variations in MRI data, we evaluate cross-domain classification accuracy within the ADNI and CALSNIC datasets by considering two scanner manufacturers' data as source domains and the remaining one as the target domain. Table 2 reflects the results by analyzing three transfer tasks from different domain combinations (source1, source2 → target): GE, Philips → Siemens; GE, Siemens → Philips; Philips, Siemens → GE. The average classification accuracy of AD vs. HC, when training and testing occur within the same domains, is approximately 90% for both the ADNI1 and ADNI2 datasets but drops to around 80% when the test/target domain is changed (w/o DA). However, by applying the proposed DA, the mean classification accuracy returns to about 89%, demonstrating that the proposed method robustly adapts the data from a different scanner to achieve a classification accuracy similar to that of the source domains. The classification accuracy of the CALSNIC datasets (ALS vs. HC) declines by approximately 13% when the test domain differs from the source domains. Nevertheless, an improvement of about 10% in the classification accuracy is regained using the proposed method. Furthermore, in Table 2, we present an ablation study focusing on the effectiveness of combining the MMD and CORAL loss functions in the network, which slightly outperforms compared to using only one. Finally, we reproduce the results of two multi-source UDA techniques, M3SDA [20] and MFSAN [31], where the proposed framework surpasses them in terms of classification accuracy.

Table 2. The cross-domain intra-study classification accuracy for the ADNI1, ADNI2, CALSNIC1 and CALSNIC2 datasets. SD: Source Domain

Dataset	Training on SD	Testing on SD	Target domain	Testing on target domain					
				w/o DA	MMD	CORAL	M3SDA	MFSAN	Proposed
ADNI1	GE+Siemens	0.90	Philips	0.80	0.83	0.86	0.82	0.82	**0.88**
	GE+Philips	0.91	Siemens	0.80	0.90	**0.91**	0.85	0.87	**0.91**
	Siemens+Philips	0.89	GE	0.81	**0.87**	0.85	0.84	0.85	**0.87**
ADNI2	GE+Siemens	0.89	Philips	0.79	**0.89**	0.87	0.86	0.87	**0.89**
	GE+Philips	0.92	Siemens	0.81	0.87	0.87	0.84	0.87	**0.88**
	Siemens+Philips	0.92	GE	0.82	0.91	0.90	0.87	0.86	**0.92**
CALSNIC1	GE+Siemens	0.75	Philips	0.56	0.64	0.62	0.63	0.64	**0.68**
	GE+Philips	0.75	Siemens	0.65	**0.77**	**0.77**	0.72	0.73	**0.77**
	Siemens+Philips	0.77	GE	0.60	0.65	0.66	0.63	0.63	**0.68**
CALSNIC2	GE+Siemens	0.77	Philips	0.60	0.73	0.72	0.71	0.70	**0.74**
	GE+Philips	0.69	Siemens	0.54	0.59	**0.65**	0.62	0.59	**0.65**
	Siemens+Philips	0.75	GE	0.68	**0.80**	0.76	0.71	0.76	**0.80**

4.3   Inter-study Validation

To validate the robustness of our findings, we analyze diverse datasets, including those from single/multiple vendors, as well as datasets containing single/multiple models from the same vendor. Table 3 shows the results, and as expected, a noticeable drop in accuracy occurs when the target data differs from the source without DA. However, the proposed framework substantially improves the accuracy after DA, with accuracy similar to that obtained in the source domains. Another ablation study examines the efficacy of addressing the target domain heterogeneity (TDH). Our approach processes the target domain based on the scanner manufacturer, producing better results than considering all samples without TDH. For datasets like AIBL and MIRIAD, our proposed method and the baseline w/o TDH exhibit the same classification accuracy as their data originated from a single manufacturer. Finally, the proposed method outperforms others in classification accuracy using the same data as ours after domain categorization.

Table 3. Inter-study classification accuracy for the ADNI, AIBL, MIRIAD and CAL-SNIC datasets. TDH: Target Domain Heterogeneity, SD: Source Domain

Training on SD	Testing on SD	Target domain	Testing on target domain						
			w/o DA	w/o TDH	MMD	CORAL	M3SDA	MFSAN	Proposed
ADNI1	0.89	ADNI2	0.81	0.83	0.88	0.89	0.87	0.87	**0.90**
		AIBL	0.75	**0.84**	0.83	0.82	0.80	0.83	**0.84**
		MIRIAD	0.78	**0.88**	**0.88**	0.85	0.85	**0.88**	**0.88**
ADNI2	0.91	ADNI1	0.79	0.84	0.87	0.87	0.84	0.85	**0.88**
		AIBL	0.74	**0.82**	**0.82**	0.82	0.80	0.81	**0.82**
		MIRIAD	0.75	**0.87**	**0.87**	0.85	0.82	**0.87**	**0.87**
CALSNIC1	0.75	CALSNIC2	0.61	0.69	0.70	**0.73**	0.68	0.69	**0.73**
CALSNIC2	0.73	CALSNIC1	0.64	0.72	0.76	0.74	0.71	0.73	**0.77**

4.4 Limitations

The second dominant factor typically observed for data variation is the scanner model. It would be interesting to analyze the results by subdividing each manufacturer's data into different models. The same vendor can reconstruct MRI data differently for various models. However, we have limited our domain consideration to the manufacturer level due to inadequate data availability for different scanner models. Moreover, the proposed architecture is task-specific and designed for classification. The network's backbone and loss functions require modification for other tasks, such as segmentation and registration.

5 Conclusion

The issue of scanner bias can negatively impact the reliability of automated analysis of MR images. The proposed framework addresses the undesirable scanner effects of multi-center MRI data and improves the consistency in classification task of such data. Most importantly, our method not only enables the pooling of data acquired by different sites within a project, but also promotes the sharing of data among different studies. Our novel strategy can substantially improve the cross-domain classification accuracy of AD/ALS patients from healthy controls. In the future, we plan to assess other MRI modalities, such as FLAIR images.

Data use declaration and acknowledgment. ADNI1, ADNI2, and AIBL neuroimaging data were collected from the ADNI portal (adni.loni.usc.edu) through a standard application process. The MIRIAD dataset was obtained from (http://miriad.drc.ion.ucl.ac.uk). This study has been supported by the Canadian Institutes of Health Research (CIHR), ALS Society of Canada, Brain Canada Foundation, Natural Sciences and Engineering Research Council of Canada (NSERC), and Prime Minister Fellowship Bangladesh.

References

1. Ackaouy, A., Courty, N., Vallée, E., Commowick, O., Barillot, C., Galassi, F.: Unsupervised domain adaptation with optimal transport in multi-site segmentation of multiple sclerosis lesions from mri data. Front. Comput. Neurosci. **14**, 19 (2020)

2. Deng, J., Dong, W., Socher, R., Li, L.J., Li, K., Fei-Fei, L.: Imagenet: a large-scale hierarchical image database. In: IEEE Conference on CVPR, pp. 248–255 (2009)
3. Dinsdale, N.K., Jenkinson, M., Namburete, A.I.L.: Unlearning scanner bias for MRI Harmonisation. In: Martel, A.L., et al. (eds.) MICCAI 2020. LNCS, vol. 12262, pp. 369–378. Springer, Cham (2020). https://doi.org/10.1007/978-3-030-59713-9_36
4. Ellis, K.A., Bush, A.I., Darby, D., De Fazio, D., Foster, J., Hudson, P., et al.: The Australian imaging, biomarkers and lifestyle (AIBL) study of aging: methodology and baseline characteristics of 1112 individuals recruited for a longitudinal study of Alzheimer's disease. Int. Psychogeriatr. **21**(4), 672–687 (2009)
5. Fortin, J.P., et al.: Harmonization of cortical thickness measurements across scanners and sites. Neuroimage **167**, 104–120 (2018)
6. Ganin, Y., Lempitsky, V.: Unsupervised domain adaptation by backpropagation. In: International Conference on Machine Learning, pp. 1180–1189. PMLR (2015)
7. Gebre, R.K., et al.: Cross-scanner harmonization methods for structural MRI may need further work: a comparison study. Neuroimage **269**, 119912 (2023)
8. Ghafoorian, M., et al.: Transfer learning for domain adaptation in MRI: application in brain lesion segmentation. In: Descoteaux, M., Maier-Hein, L., Franz, A., Jannin, P., Collins, D.L., Duchesne, S. (eds.) MICCAI 2017. LNCS, vol. 10435, pp. 516–524. Springer, Cham (2017). https://doi.org/10.1007/978-3-319-66179-7_59
9. Girshick, R.: Fast r-cnn. In: Proceedings of the IEEE International Conference on Computer Vision, pp. 1440–1448 (2015)
10. He, K., Zhang, X., Ren, S., Sun, J.: Deep residual learning for image recognition. In: Proceedings of the IEEE Conference on CVPR, pp. 770–778 (2016)
11. Jack, C.R., Jr., et al.: The Alzheimer's disease neuroimaging initiative (ADNI): MRI methods. J. Magn. Reson. Imaging **27**(4), 685–691 (2008)
12. Kalra, S., et al.: The canadian als neuroimaging consortium (calsnic)-a multicentre platform for standardized imaging and clinical studies in ALS. MedRxiv (2020)
13. Kushol, R., Luk, C.C., Dey, A., Benatar, M., Briemberg, H., et al.: Sf2former: amyotrophic lateral sclerosis identification from multi-center MRI data using spatial and frequency fusion transformer. Comput. Med. Imaging Graph. **108**, 102279 (2023)
14. Kushol, R., Masoumzadeh, A., Huo, D., Kalra, S., Yang, Y.H.: Addformer: Alzheimer's disease detection from structural mri using fusion transformer. In: IEEE 19th International Symposium on Biomedical Imaging, pp. 1–5. IEEE (2022)
15. Liu, M., et al.: Style transfer using generative adversarial networks for multi-site MRI Harmonization. In: de Bruijne, M., et al. (eds.) MICCAI 2021. LNCS, vol. 12903, pp. 313–322. Springer, Cham (2021). https://doi.org/10.1007/978-3-030-87199-4_30
16. Van der Maaten, L., Hinton, G.: Visualizing data using t-SNE. J. Mach. Learn. Res. **9**(11), 2579–2605 (2008)
17. Malone, I.B., et al.: Miriad-public release of a multiple time point Alzheimer's MR imaging dataset. Neuroimage **70**, 33–36 (2013)
18. Orbes-Arteaga, M., et al.: Multi-domain adaptation in brain MRI through paired consistency and adversarial learning. In: Wang, Q., et al. (eds.) DART/MIL3ID -2019. LNCS, vol. 11795, pp. 54–62. Springer, Cham (2019). https://doi.org/10.1007/978-3-030-33391-1_7
19. Panfilov, E., Tiulpin, A., Klein, S., Nieminen, M.T., Saarakkala, S.: Improving robustness of deep learning based knee MRI segmentation: mixup and adversarial domain adaptation. In: Proceedings of the IEEE/CVF International Conference on Computer Vision Workshops, pp. 0–0 (2019)

20. Peng, X., Bai, Q., Xia, X., Huang, Z., Saenko, K., Wang, B.: Moment matching for multi-source domain adaptation. In: Proceedings of the IEEE/CVF International Conference on Computer Vision, pp. 1406–1415 (2019)
21. Quinonero-Candela, J., Sugiyama, M., Schwaighofer, A., Lawrence, N.D.: Dataset Shift in Machine Learning. MIT Press, Cambridge (2008)
22. Sadri, A.R., et al.: MRQY-an open-source tool for quality control of MR imaging data. Med. Phys. **47**(12), 6029–6038 (2020)
23. Sun, B., Saenko, K.: Deep CORAL: correlation alignment for deep domain adaptation. In: Hua, G., Jégou, H. (eds.) ECCV 2016. LNCS, vol. 9915, pp. 443–450. Springer, Cham (2016). https://doi.org/10.1007/978-3-319-49409-8_35
24. Tian, D., et al.: A deep learning-based multisite neuroimage harmonization framework established with a traveling-subject dataset. NeuroImage **257**, 119297 (2022)
25. Wachinger, C., Reuter, M., Initiative, A.D.N., et al.: Domain adaptation for Alzheimer's disease diagnostics. Neuroimage **139**, 470–479 (2016)
26. Wang, R., Chaudhari, P., Davatzikos, C.: Embracing the disharmony in medical imaging: a simple and effective framework for domain adaptation. Med. Image Anal. **76**, 102309 (2022)
27. Wen, J., Thibeau-Sutre, E., Diaz-Melo, M., Samper-González, J., Routier, A., et al.: Convolutional neural networks for classification of Alzheimer's disease: overview and reproducible evaluation. Med. Image Anal. **63**, 101694 (2020)
28. Wolleb, J., et al.: Learn to ignore: domain adaptation for multi-site MRI analysis. In: Wang, L., Dou, Q., Fletcher, P.T., Speidel, S., Li, S. (eds.) MICCAI 2022. LNCS, vol. 13437, pp. 725–735. Springer, Cham (2022)
29. Yagis, E., et al.: Effect of data leakage in brain MRI classification using 2D convolutional neural networks. Sci. Rep. **11**(1), 1–13 (2021)
30. Zeng, L.L., et al.: Gradient matching federated domain adaptation for brain image classification. IEEE Trans. Neural Networks Learn. Syst. (2022)
31. Zhu, Y., Zhuang, F., Wang, D.: Aligning domain-specific distribution and classifier for cross-domain classification from multiple sources. In: Proceedings of the AAAI Conference on Artificial Intelligence, vol. 33, pp. 5989–5996 (2019)

MultiVT: Multiple-Task Framework for Dentistry

Edoardo Mello Rella[1][✉], Ajad Chhatkuli[1], Ender Konukoglu[1], and Luc Van Gool[1,2]

[1] Computer Vision Lab, ETH Zurich, Zürich, Switzerland
edoardo.mello-rella@vision.ee.ethz.ch
[2] VISICS, KU Leuven, Leuven, Belgium

Abstract. Current image understanding methods on dental data are often trained end-to-end on inputs and labels, with focus on using state-of-the-art neural architectures. Such approaches, however, typically ignore domain specific peculiarities and lack the ability to generalize outside their training dataset. We observe that, in RGB images, teeth display a weak or unremarkable texture while exhibiting strong boundaries; similarly, in panoramic radiographs root tips and crowns are generally visible, while other parts of the teeth appear blurry. In their respective image type, these features are robust to the domain shift given by different sources or acquisition tools. Therefore, we formulate a method which we call MultiVT, able to leverage these low level image features to achieve results with good domain generalization properties. We demonstrate with experiments that, by focusing on such domain-robust features, we can achieve better segmentation and detection results. Additionally, MultiVT improves generalization capabilities without applying domain adaptive techniques - a characteristic which renders our method suitable for use in real-world applications.

Keywords: Dental Image Understanding · Domain Generalization · Instance Segmentation · Object Detection

1 Introduction

In recent years, deep learning has shown excellent capability in solving problems that traditionally required human intelligence, ranging from image understanding [16, 21, 27], natural language processing [4, 5], machine control [1, 18] as well as many others [10, 15, 22]. Arguably, one of the innovations that brought the most advances has been the use of convolutional neural networks (CNN) [13, 16, 27], which has revolutionised image understanding. With the advancement of learning methods, similar progress has been observed in the use of machine intelligence

Supplementary Information The online version contains supplementary material available at https://doi.org/10.1007/978-3-031-45857-6_2.

in the medical field [8,30], in which diagnosis and intervention frequently involve the acquisition and analysis of images. Thus, it appears logical to ask whether computer vision can facilitate or partially solve medical image understanding. More specifically, dentistry makes use of image acquisitions for diagnosis and planning of interventions, at almost every step of a patient's treatment, and can benefit from the advances in computer vision [6]. Despite the clear potential, the amount of research work in the field is relatively scarce.

An additional challenge in using computer vision methods for the medical domains is developing methods that are robust and able to generalize in the presence of domain shifts [9,11,17,19]. This is especially significant in clinical applications, since stable performances are of paramount importance for the utilization in a clinical setting [28]. Furthermore, domain shift is extremely common and of key importance in the medical field, as it arises from different conditions or from the use of heterogeneous equipment and parameterizations [23] in the image acquisition process. Despite being subtle and, at times, indiscernible to the human observer, such differences can significantly hamper the accuracy of a learned method [7,23]. Therefore, it is of vital importance to design robust algorithms while noting that we may not always recognize domain shifts in the input data.

In this work, we focus on the segmentation and detection tasks on dental data, with the specific aim of building a method that is more robust to domain shift, without using adaptation steps. We develop a prediction framework, which we call MultiVT, which uses the vector transform (VT) [25] to exploit low level image features, in order to improve performance and generalization. By incorporating in the prediction structure VT, a powerful generic surface representation that consists of the direction towards the closest surface, MultiVT can focus on the salient parts of objects, like visible lines or points. When these salient parts are carefully selected to retain similar appearance across domains, they constitute strong and robust priors. This can improve accuracy and generalization in the identification and localization of the desired objects. For example, in RGB images, teeth often lack texture; thus exploiting the boundary information helps to improve performance. Similarly, by concentrating on specific teeth parts, MultiVT outperforms state-of-the-art methods on detecting and numbering teeth in panoramic radiographs, with rare false positives. An even more interesting result is that in both cases MultiVT can generalize to different image types with almost no visible performance degradation, while traditional methods suffer significant performance drops.

In summary, our contributions are threefold:

- We propose MultiVT, a framework based on VT, which is able to incorporate low level image details and use them across multiple tasks.
- We show that MultiVT can be applied to the instance segmentation and object detection tasks, improving their performance over previous state-of-the-art methods.
- We demonstrate that, unlike other methods, MultiVT can generalize to different domains with no adaptation techniques, by focusing on low level features of teeth that show little variation across image types.

14   E. Mello Rella et al.

2 Method

MultiVT is a prediction framework that uses VT [25] to represent low level image details and integrates them to formulate accurate and robust predictions in the tasks of segmentation and detection. In this section, we first review the VT representation and expand its capability to represent different structures in an image. We then describe how it can be applied to the tasks of instance segmentation and object detection. Finally, we explain the neural network architecture used for MultiVT, together with the training and post-processing techniques (Fig. 1).

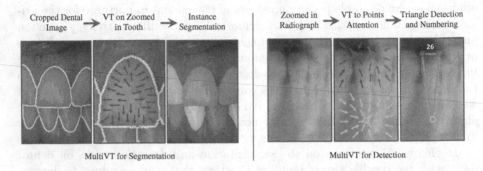

Fig. 1. Method overview: We depict MultiVT emphasizing the representation used for segmentation and detection. **On the left**, we visualize MultiVT for instance segmentation. We first show a few teeth with their boundaries highlighted in white. Then, we visually represent VT by zooming in on a single tooth; for clarity, VT is displayed only inside one tooth, while, in practice, it is predicted at every pixel. Finally, we exhibit the instance segmentation obtained by using MultiVT. **On the right**, we show our novel approach for point representation with VT, providing tooth detection with numbering. The three different colors in the tooth vertices correspond to the three point types: they are shown on the same image, although, in practice, they are predicted in three different channels.

2.1 MultiVT Representations

VT as a boundary representation [25] is defined in each pixel as the direction vector v pointing towards the closest boundary point. We extend the definition to represent any relevant element in an image, either lines or points. Given a set of interest points Ω - *e.g.* points in a line or isolated - we define v at any pixel location x as

$$v = -\frac{x - \hat{x}_S}{\|x - \hat{x}_S\|} \quad \text{with} \quad \hat{x}_S = \operatorname*{argmin}_{x_S \in \Omega} \|x - x_S\|$$

or, in simpler terms, as the direction towards the closest among the interest points. If $x \in \Omega$, the equation is not well defined, but in practice it is solved by assigning the value v of any of the closest pixels $x \notin \Omega$ arbitrarily. Once a distinctive feature in a class of images is identified, we can train a neural network to predict the

VT representation of such feature. As an example, given that teeth boundaries are a visible feature in RGB images, we train a network to predict at each pixel the direction towards the closest boundary point. We name our method MultiVT, since it utilizes the definition of VT in tackling multiple tasks.

MultiVT for Instance Segmentation. As the first problem, we apply MultiVT to instance segmentation of teeth images, the task that consists of identifying masks of individual teeth. Since teeth typically lack texture in RGB images, but exhibit sharp boundaries, the boundaries can be used as a strong prior for prediction. Additionally, the appearance of boundaries remain robust under variations of light, multiple image acquisitions set-ups (*e.g.* w/wo cheek retractor), and different view angles. Therefore, MultiVT for instance segmentation incorporates the VT representation of teeth boundaries; by doing so, the network can learn features that remain similar across teeth types and different views and that are useful to generate accurate instance masks.

In practice, MultiVT for instance segmentation is composed of two parts which share the same backbone; an instance segmentation network similar to [3] which generates proposal instance masks and the VT prediction which refines the aforementioned masks. The instance segmentation part of MultiVT uses three prediction heads, which output semantic segmentation, instance center, and instance offset. Semantic segmentation is the mask that discerns the teeth from the background. Center and offset are, respectively, a heatmap comprised of Gaussians located at each instance center, and the 2-D vector offset of each pixel inside an instance to its center. The VT part has a single prediction head which outputs the direction towards the closest boundary, and is predicted in 8 discrete bins as a classification problem. This prediction mode was preferred over the angle component regression based on their empirical performance.

The generation of instance segmentation masks is executed by grouping pixels based on the predictions of the instance segmentation part of MultiVT while ensuring that the masks are consistent with the VT part. To that end, we first refine the output of the instance offset and semantic segmentation head using the VT predictions. We perform this iteratively: for each pixel at location x_{start}, we follow the path opposite to the direction of the VT prediction for 60 iterations - *i.e.* going away from the boundary. Each time, the direction is chosen from the VT of the pixel closest to the current position, which changes after each step. After the iterations, each pixel will have reached a final position x_{end}. For each pixel inside an instance, the path approaches the center of the instance; pixel outside instances are moved away from their boundaries. Finally, we assign a new offset to the original pixel at location x_{start} as the sum of the net vector path $(x_{end} - x_{start})$ and the offset at the end-of-path x_{end}. Similarly, the semantic label is re-assigned to the original pixel from the end-of-path - *i.e.* the pixel at position x_{start} has the semantic label that was predicted at position x_{end}. In practice, this results in all pixels inside a boundary having homogeneous semantics and offset values consistent to the same center position. The output instance segmentation is then obtained by using the instance centers and the updated semantic and offset, similarly to [3]. In the pipeline described, the use of VT ensures that every pixel inside a boundary is assigned to the same instance, avoiding problems such as over-segmentation or low quality boundaries.

MultiVT for Teeth Detection and Numbering. In object detection, one identifies the locations of a set of objects in an image and assigns them a semantic label. In our case, from a panoramic radiograph, we detect teeth positions and assign each detection their tooth number. To understand the tooth orientation besides its position, we formulate the detection using bounding triangles, with one vertex at the center of the tips of the roots, and two vertices on the two sides of the crowns. Observing that these types of points are visible in multiple types of radiographs and maintain consistent appearance, MultiVT uses VT to detect the vertices of the triangles by representing them as points - *i.e.* VT is created using as set of interest points Ω the vertices of the detection triangles. This allows one to localize predictions based on visual cues that remain similar across teeth and different domains, and to minimize the number of false positives.

Similarly to instance segmentation, MultiVT for teeth detection and numbering has two parts; a semantic and a VT part. The VT part has a single head that predicts the direction towards the closest of the vertices of every bounding triangle in three separate channels; one channel per vertex type - *i.e.* one channel predicts VT only for the points on the tips of the roots of each tooth, one other only for the points on one side of the crowns, *etc.* Again, VT is parameterized as 8 discrete bins and approached as a classification problem. The semantic part has a single segmentation head, which executes the teeth numbering by predicting three separate semantic masks - one for each vertex type similarly to the VT part - with the classes that are the teeth numbers.

During the inference process, we extract the vertices of the bounding triangles from VT and assign them their labels with the semantic part. We obtain the vertices by i) computing the discrete divergence on the predicted VT field, ii) applying non-maximum suppression and, iii) by accepting the points with a divergence value lower than -2. Then, each vertex is assigned the tooth number predicted at its pose in the corresponding channel of the semantic head. Even though it is possible to have the same label assigned to more or less than three points, we observe that in practice this happens only in rare cases (less than 1% of the detections). In these rare cases, the problem can be easily solved by removing the additional points at random. We observe that such a simple postprocessing method is as effective as more advanced heuristics given proximity of the predicted points in these cases. In presence of only two valid points in a tooth, the missing point is added to mimic the average triangle shape in the image. When two points are missing, the prediction is discarded.

2.2   MultiVT Network and Training

To best exploit the high resolution details of the objects represented with VT, we build the overall architecture to predict at high resolution. We use HRNet

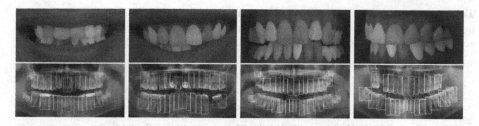

Fig. 2. Qualitative results of MultiVT on instance segmentation (first row) and teeth detection and numbering (second row). In the first row the first two images are from Smiling Images dataset and the last two from Cheek Retractor Images dataset. In the second row the first two images are from Low Res PR dataset and the last two from High Res PR dataset. For visualization purposes, the detected teeth are represented with bounding boxes taken around the predicted bounding triangles. In red are the predicted boxes and numbering and in green the corresponding ground truth. (Color figure online)

[29] backbone and bilinearly up-sample its output to match the input resolution. The output features are then processed by a set of 3-layered convolutional heads, whose number and output size depend on the application.

We train the network parameters by jointly optimizing the output of each network head. Each model is trained using Adam optimizer [14] for 100 000 iterations with a batch size of 32. We use poly learning scheduler with a starting learning rate of 0.001 after an initial warm-up phase of 1 000 iterations. For each task, the input images are augmented with image resizing, random cropping and random horizontal flipping. We assume the mouth to be symmetrical and swap the target teeth labels, when images are flipped.

The instance segmentation loss ℓ_{IS} is formulated as:

$$\ell_{IS} = \ell_s + \alpha\,\ell_c + \beta\,\ell_o + \gamma\,\ell_{VTc}$$

Here, the semantic loss ℓ_s and the loss on VT ℓ_{VT} are CrossEntropy losses. Additionally, the loss on the center ℓ_c is the Mean Squared Error and the loss on the offset ℓ_o is the L1 loss. The weights are set to $\alpha = 200$, $\beta = 0.01$, and $\gamma = 5$ and are selected after running a grid search with $\alpha \in [50, 100, 200, 400]$, $\beta \in [0.005, 0.01, 0.02, 0.04]$, $\gamma \in [1, 5, 10, 20]$ in the validation set.

The detection and numbering loss ℓ_{DN} is composed of a VT component ℓ_{VT} and a semantic component ℓ_s which are CrossEntropy losses.

$$\ell_{DN} = \ell_s + \delta\,\ell_{VTr}$$

We set $\delta = 10$ with a grid search for $\delta \in [1, 5, 10, 20, 40]$ in the validation set.

3   Experiments

First, we describe the data setup and the metrics. Then, we discuss the methods used as baselines and show how they compare to MultiVT on instance segmentation and object detection.

Table 1. Quantitative comparison on instance segmentation. Each metric is shown in percentage points and we show the mean and the standard deviation of the results across 6 different experiments.

Method	Smiling Images		Cheek Retractor Images	
	mAP	mAP@50	mAP	mAP@50
Panoptic DeepLab [3]	53.1/0.6	78.9/0.9	46.4/1.7	69.9/2.8
Mask-RCNN [12]	49.6/1.2	75.3/1.5	42.9/1.9	64.0/3.0
DeepLabV3+ [2]	50.6/0.3	76.0/0.5	36.2/2.5	57.8/3.8
MultiVT	**53.6**/0.4	**79.6**/0.4	**47.3**/0.6	**73.6**/1.3

3.1 Datasets and Metrics

With the aim of evaluating the domain generalization capabilities of the methods together with their accuracy, the datasets are divided into two parts. For instance segmentation, the first part, which we call Smiling Images, is composed of images taken while patients are smiling and is divided in around 7 000 images for training, 1 000 for validation and 2 000 for testing. The second part, which we call Cheek Retractor Images, is composed of patients wearing a cheek retractor and taken from either a frontal or 45° lateral position. The second part is not used for training and is divided into about 2 000 images for validation and 4 000 for testing. Examples of the two types of images can be seen in Fig. 2. Similarly, for teeth detection and numbering, the first part, called Low Resolution Panoramic Radiographs (Low Res PR), is used for training (700 images) as well as for validation and testing (100 and 200 images), while the second part, called High Resolution Panoramic Radiographs (High Res PR) is used only for validation (300 images) and testing (700 images). The two datasets are collected from different practitioners and Low Res PR has images with an average resolution of 320×480 while High Res PR 1200×1800. All datasets are proprietary anonymized from Align Technology Switzerland GmbH, internal use only, no license specified.

As typically done in instance segmentation, we report the results as mean average precision (mAP) and mAP with overlapping threshold of 50% (mAP@50). On detection and numbering, we use the common mean bounding box intersection over union (mIoU). We also evaluate the Precision, Recall and F-Score with a 50% box overlap as well as Precision, Recall and F-Score on the accuracy of correctly recognizing missing teeth (we call them MT Precision, MT Recall and MT F-Score).

3.2 Instance Segmentation

On Instance segmentation we compare MultiVT to the popular instance segmentation methods Panoptic DeepLab [3] and Mask-RCNN [12]; additionally, we compare to the semantic segmentation method DeepLabV3+ [2] by providing as training input the teeth masks together with their numbering. The results

Table 2. Quantitative comparison on the teeth detection and numbering. Each metric is shown in percentage points and we show the mean and the standard deviation of the results across 6 different experiments.

Method	mIoU	Precision	Recall	F-score	MT Precision	MT Recall	MT F-score
Low Res PR							
Faster RCNN [26]	71.8/0.1	90.2/0.1	**97.9**/0.1	93.7/0.1	**97.8**/0.5	78.8/0.7	87.3/0.3
SSD [20]	74.7/0.1	93.4/0.2	96.6/0.2	95.0/0.2	95.3/0.6	83.4/0.8	89.0/0.6
YOLO [24]	75.3/0.1	93.6/0.2	97.0/0.2	95.3/0.2	96.0/0.5	85.7/0.8	90.6/0.6
MultiVT	**77.7**/0.1	**96.2**/0.2	96.3/0.2	**96.2**/0.2	92.8/1.2	**92.1**/0.6	**92.4**/0.9
High Res PR							
Faster RCNN [26]	58.1/0.2	75.3/0.3	92.6/0.2	82.7/0.2	85.2/0.3	57.9/1.3	69.0/1.0
SSD [20]	62.8/0.3	78.8/0.3	94.0/0.3	85.7/0.3	87.9/0.5	62.1/0.9	72.8./0.6
YOLO [24]	64.3/0.5	79.2/0.5	**94.8**/0.4	86.3/0.4	**89.0**/0.7	63.6/1.1	74.2/0.9
MultiVT	**73.9**/1.2	**95.0**/0.7	94.5/0.8	**94.7**/0.8	86.8/1.7	**93.1**/0.5	**89.8**/1.0

in Table 1 show that VT can improve performance on the same domain as well as across domains. The difference in performance increases when the inference domain is different from that of the training, specifically it goes from 0.9% to 5.3% compared to the second best method on mAP@50. This demonstrates the ability of MultiVT to improve over previous methods and to generalize better across domains without adaptation techniques. Top row in Fig. 2 further shows the ability of MultiVT to predict on the two domains with almost no performance degradation.

3.3 Teeth Detection and Numbering

For teeth detection and numbering, we compare MultiVT to the most widely used object detection methods, Faster RCNN [26], SSD [20], and YOLO [24], and adapt them to predict bounding triangles. Results in Table 2 show that MultiVT can outperform previous state of the art on most metrics. Traditional methods only achieve a better Recall and a better MT Precision because they suffer from the presence of numerous false positives (on average 3.4 or more per image against 1.1 of MultiVT). Overall, however, MultiVT outperforms Faster RCNN, SSD and YOLO in terms of F-score and MT F-Score, which consider both Precision and Recall. Similarly to instance segmentation, the relative improvement on mIoU compared to YOLO goes from 3.2% to 14.9% when moving from Low Res PR to High Res PR, which shows the higher generalization capabilities of MultiVT. We show qualitative results in Fig. 2.

4 Conclusions

In this work, we extended and adapted VT, a recently proposed representation, for tackling different relevant tasks in the field of dentistry. More specifically, we proposed MultiVT, a framework for solving multiple dental image understanding problems with potential clinical applications. We showed that MultiVT achieves performance above current state of the art models while improving also in terms of generalization. We hope the results we showed in this work can pave the way for further exploration and automation in the dental domain.

Acknowledgements. This research was funded by Align Technology Switzerland GmbH (project AlignTech-ETH). Research was also partially funded by VIVO Collaboration Project on Real-time scene reconstruction.

References

1. Burgner-Kahrs, J., Rucker, D.C., Choset, H.: Continuum robots for medical applications: a survey. IEEE Trans. Robot. **31**, 1261–1280 (2015)
2. Chen, L.C., Zhu, Y., Papandreou, G., Schroff, F., Adam, H.: Encoder-decoder with atrous separable convolution for semantic image segmentation. In: Proceedings of the European Conference on Computer Vision (ECCV) (2018)
3. Cheng, B., et al.: Panoptic-DeepLab: a simple, strong, and fast baseline for bottom-up panoptic segmentation. In: IEEE Conference on Computer Vision and Pattern Recognition (2020)
4. Collobert, R., Weston, J.: A unified architecture for natural language processing: Deep neural networks with multitask learning. In: Proceedings of the 25th International Conference on Machine Learning (2008)
5. Devlin, J., Chang, M.W., Lee, K., Toutanova, K.: BERT: pre-training of deep bidirectional transformers for language understanding. In: Proceedings of the 2019 Conference of the North American Chapter of the Association for Computational Linguistics: Human Language Technologies, Volume 1 (Long and Short Papers) (2019)
6. Schwendicke, F., Samek, W., Krois, J.: Artificial intelligence in dentistry: chances and challenges. J. Dent. Res. **99**, 769–774 (2020)
7. Finlayson, S.G., Bowers, J.D., Ito, J., Zittrain, J.L., Beam, A.L., Kohane, I.S.: Adversarial attacks on medical machine learning. Science **363**(6433), 1287–1289 (2019)
8. Frid-Adar, M., Diamant, I., Klang, E., Amitai, M., Goldberger, J., Greenspan, H.: Gan-based synthetic medical image augmentation for increased CNN performance in liver lesion classification. Neurocomputing **321**, 321–331 (2018)
9. Ganin, Y., Lempitsky, V.: Unsupervised domain adaptation by backpropagation. In: Proceedings of the 32nd International Conference on Machine Learning (2015)
10. Goodfellow, I., et al.: Generative adversarial networks. Commun. ACM **63**, 139–144 (2020)
11. Guan, H., Liu, M.: Domain adaptation for medical image analysis: a survey. IEEE Trans. Biomed. Eng. **69**, 1173–1185 (2022)
12. He, K., Gkioxari, G., Dollar, P., Girshick, R.: Mask R-CNN. In: Proceedings of the IEEE International Conference on Computer Vision (ICCV) (2017)
13. He, K., Zhang, X., Ren, S., Sun, J.: Deep residual learning for image recognition. In: Proceedings of the IEEE Conference on Computer Vision and Pattern Recognition, pp. 770–778 (2016)
14. Kingma, D.P., Ba, J.: Adam: a method for stochastic optimization. In: 3rd International Conference on Learning Representations, ICLR (2015)
15. Koprinska, I., Carrato, S.: Temporal video segmentation: a survey. In: Signal Processing: Image Communication (2001)

16. Krizhevsky, A., Sutskever, I., Hinton, G.E.: Imagenet classification with deep convolutional neural networks. In: Advances in Neural Information Processing Systems (2012)
17. Li, D., Yang, Y., Song, Y.Z., Hospedales, T.M.: Deeper, broader and artier domain generalization. In: Proceedings of the IEEE International Conference on Computer Vision (ICCV) (2017)
18. Lillicrap, T.P., et al.: Continuous control with deep reinforcement learning. In: 4th International Conference on Learning Representations, ICLR (2016)
19. Liu, Q., Chen, C., Qin, J., Dou, Q., Heng, P.A.: FEDDG: federated domain generalization on medical image segmentation via episodic learning in continuous frequency space. In: Proceedings of the IEEE/CVF Conference on Computer Vision and Pattern Recognition (CVPR) (2021)
20. Liu, W., et al.: SSD: single shot multibox detector. In: Leibe, B., Matas, J., Sebe, N., Welling, M. (eds.) ECCV 2016. LNCS, vol. 9905, pp. 21–37. Springer, Cham (2016). https://doi.org/10.1007/978-3-319-46448-0_2
21. Long, J., Shelhamer, E., Darrell, T.: Fully convolutional networks for semantic segmentation. In: Proceedings of the IEEE Conference on Computer Vision and Pattern Recognition (CVPR) (2015)
22. Mildenhall, B., Srinivasan, P.P., Tancik, M., Barron, J.T., Ramamoorthi, R., Ng, R.: Nerf: representing scenes as neural radiance fields for view synthesis. Commun. ACM **65**, 99–106 (2021)
23. Pooch, E.H.P., Ballester, P., Barros, R.C.: Can we trust deep learning based diagnosis? The impact of domain shift in chest radiograph classification (2020)
24. Redmon, J., Divvala, S., Girshick, R., Farhadi, A.: You only look once: unified, real-time object detection. In: 2016 IEEE Conference on Computer Vision and Pattern Recognition (CVPR), pp. 779–788 (2016). https://doi.org/10.1109/CVPR.2016.91
25. Rella, E.M., Chhatkuli, A., Liu, Y., Konukoglu, E., Gool, L.V.: Zero pixel directional boundary by vector transform. In: International Conference on Learning Representations (2022)
26. Ren, S., He, K., Girshick, R., Sun, J.: Faster R-CNN: towards real-time object detection with region proposal networks. In: Advances in Neural Information Processing Systems (2015)
27. Ronneberger, O., Fischer, P., Brox, T.: U-net: convolutional networks for biomedical image segmentation. In: Navab, N., Hornegger, J., Wells, W.M., Frangi, A.F. (eds.) MICCAI 2015, Part III. LNCS, vol. 9351, pp. 234–241. Springer, Cham (2015). https://doi.org/10.1007/978-3-319-24574-4_28
28. Schwendicke, F., et al.: Artificial intelligence in dental research: checklist for authors, reviewers, readers. J. Dent. **10**, 103610 (2021)
29. Wang, J., et al.: Deep high-resolution representation learning for visual recognition. IEEE Trans. Pattern Anal. Mach. Intelli. **43**, 3349–3364 (2020)
30. Yadav, S.S., Jadhav, S.M.: Deep convolutional neural network based medical image classification for disease diagnosis. J. Big Data **6**, 1–18 (2019)

Black-Box Unsupervised Domain Adaptation for Medical Image Segmentation

Satoshi Kondo[✉][ID]

Muroran Institute of Technology, Hokkaido, Japan
kondo@muroran-it.ac.jp

Abstract. Unsupervised Domain Adaptation (UDA) is one of the key technologies to solve the problem of obtaining ground truth labels needed for supervised learning. In general, UDA assumes that information about the source model, such as its architecture and weights, and all samples from the source domains are available when a target domain model is trained. However, this is not a realistic assumption in applications where privacy and white-box attacks are a concern, or where the model is only be accessible through an API. To overcome this limitation, UDA without source model information and source data, called Black-Box Unsupervised Domain Adaptation (BBUDA), has recently been proposed. Here, we propose an improved BBUDA method for medical image segmentation. Our main contribution is the introduction of a mean teacher algorithm during the training of the target domain model. We conduct experiments on datasets containing different types of source-target domain combinations to demonstrate the versatility and robustness of our method. We confirm that our method outperforms the state-of-the-art on all datasets.

Keywords: Unsupervised domain adaptation · Image segmentation · Black-box model · Self-supervised learning

1 Introduction

Image segmentation plays an important role in medical image analysis. In recent years, deep learning has been widely used for medical image segmentation. The performance of segmentation has been greatly improved with supervised deep learning for a variety of imaging modalities and target organs [21]. However, there are some problems with supervised deep learning for segmentation tasks. The first is that supervised learning requires a ground truth label for each pixel. This is a time-consuming task, and it is also difficult to obtain a large amount of labeled data for medical images. The second problem is that a model trained on a dataset for one modality or organ is difficult to apply to other modalities or organs, as the performance of the model typically degrades.

© The Author(s), under exclusive license to Springer Nature Switzerland AG 2024
L. Koch et al. (Eds.): DART 2023, LNCS 14293, pp. 22–30, 2024.
https://doi.org/10.1007/978-3-031-45857-6_3

Unsupervised Domain Adaptation (UDA) is one of the solutions to these problems [6,20]. UDA is a type of domain adaptation that uses labeled data from the source domain and unlabeled data from the target domain. In general, UDA assumes that information about the source model, such as its architecture and weights, and all samples from the source and target domains are available during the training process of the target model. The models under this assumption are called "white-box unsupervised domain adaptation". However, this is not a realistic assumption in applications where privacy is an issue, e.g., when source and target data come from different clinical sites and white-box attacks [22] are a concern. And this is impossible if the model is only be accessible through an API, e.g. to a web service. To overcome this limitation, UDA without source model information and source data, called Black-Box Unsupervised Domain Adaptation (BBUDA), has recently been proposed.

BBUDA methods are mainly proposed for image classification tasks [9,10, 16,23]. On the other hand, Liu et al. propose a BBUDA method for a medical image segmentation task [12]. This is the only prior art for image segmentation in the BBUDA setting that we could find. Their method combines a self-training scheme [7] and self-entropy minimization [5]. In the self-training scheme, they construct pseudo-segmentation labels for training in the target domain. The pseudo labels are obtained by mixing the predictions of the source and target models. The contribution of the source model prediction is adjusted to be large at the beginning of training and smaller in the later training epochs. The KL divergence between the pseudo labels and the target model predictions is the loss function for self-training. In self-entropy minimization, the averaged entropy of the pixel-wise softmax prediction obtained by the target model is the loss function. A weighted sum of two loss functions is minimized during training of the target model. The method is evaluated using 3D multi-modal magnetic resonance imaging of the brain.

In the method proposed by Liu et al. [12], the pseudo-labels are not reliable if there is a gap between the source and target domains. In the case of white-box unsupervised domain adaptation, a target domain model can be initialized with the weights of the source domain model, but in the BBUDA setting, a target domain model is initialized with random values or pre-trained weights, e.g., pre-trained using the ImageNet dataset. Therefore, it is important to stabilize the training procedure of the target domain model in the BBUDA setting. In addition, Liu et al. [12] only use statistical loss function, i.e., KL divergence and self-entropy loss, but morphological information needs to be considered to improve the performance of the segmentation task.

We propose an improved BBUDA method for medical image segmentation. Our proposed method uses a similar framework with [12] to address BBUDA image segmentation tasks, but we extend their method in two ways as mentioned below. First, we introduce a Mean Teacher (MT) algorithm [19] to stabilize the training during the training of the target domain model. We also introduce a Dice loss to account for morphological information during the training of the target domain model.

Besides the technical aspects, in [12] the evaluation is performed on only one data set, i.e., brain MRI images. In contrast, we prepare two different types of source-target combinations, i.e., brain MRI images and color fundus images, for the evaluation of our method to confirm the versatility and robustness of our method.

We summarize our contributions as follows:

1. To the best of our knowledge, this is the first attempt to introduce the MT algorithm to BBUDA for medical image segmentation, in order to stabilize the training during the training of the target domain model.
2. We compare our method with conventional methods using datasets with different types of source-target domain combinations and confirm the performance of image segmentation in BBUDA.

2   Proposed Method

We address the BBUDA task, which is the UDA task without access to both source model information and images in the domain adaptation phase, for medical image segmentation. We are given a segmentation model f_s for the source domain D_s, which is trained by using n_s labeled samples $\{x_s^i, y_s^i\}_{i=1}^{n_s}$ from the source domain, where x_s^i is a source image and y_s^i is the corresponding ground truth with a K-class label for each pixel in x_s^i. We are also given n_t unlabeled samples $\{x_t^i\}_{i=1}^{n_t}$ from the target domain D_t. The goal of BBUDA for image segmentation is to train a segmentation model f_t for the target domain and predict the segmentation labels $\{y_t^i\}_{i=1}^{n_t}$ in the target domain, without access to $\{x_s^i, y_s^i\}_{i=1}^{n_s}$ and the interior of f_s in the domain adaptation phase.

Figure 1 shows the block diagram of the domain adaptation phase. We have three segmentation networks. The first one is the source domain model f_s, and it is black box and fixed during the domain adaptation phase. The second and the third are the student and the MT target domain models f_t, respectively. An input image x_t from the target domain is processed by all three networks. A pseudo label y_t' is obtained by weighted averaging of the source and target domain predictions, i.e., $f_s(x_t)$ and $f_t(x_t)$, as in Eq. (1).

$$y_t' = \alpha f_s(x_t) + (1 - \alpha)f_t(x_t; \theta),\tag{1}$$

where α is a weight parameter and θ is network parameters of the student target model. α is set to 1 at the start of training and is gradually reduced to 0 at each epoch. Thus, α adjusts the contribution of the source model prediction to be large at the beginning of training and smaller in the later training epochs.

Our loss function has four terms. The first and fourth terms are losses between the pseudo-labels and the predictions, and the first term is the KL divergence loss and the fourth term is the Dice loss, as in Eqs. (2) and (3), respectively.

$$L_K = D_{KL}(f_t(x_t; \theta)||y_t'),\tag{2}$$

$$L_D = 1 - \mathrm{DICE}(f_t(x_t; \theta), y_t'),\tag{3}$$

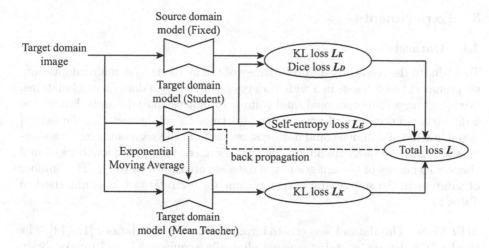

Fig. 1. Schematic diagram of network architectures and loss functions at the domain adaptation phase in our proposed method.

where D_{KL} is the KL divergence and DICE() is the Dice coefficient between pseudo-labels and predicted labels. The third term is the self-entropy loss for prediction by the student target model as in Eq. (4).

$$L_E = -f_t(x_t; \theta) \log f_t(x_t; \theta). \tag{4}$$

The second term is the similarity between the predictions of the student and MT target models.

$$L'_K = D_{KL}(f_t(x_t; \theta) \| f_t(x_t; \theta')), \tag{5}$$

where θ' is the network parameters of the MT target model. The total loss function L is the sum of these four losses.

$$L = L_K + L'_K + \beta L_E + (1 - \beta) L_D, \tag{6}$$

where β is a weighting parameter. β is set to 1 at the beginning of training and is gradually reduced to 0 at each epoch. Thus, β adjusts the contribution of the self-entropy loss L_E to be large and the Dice loss L_D to be small at the beginning of training, and smaller and larger in the later training epochs, respectively. The self-entropy loss L_E is a statistical measure of the predictions. And the Dice loss L_D is a measure of the morphological difference between the pseudo-labels and the predictions. Therefore, the statistical distribution of the predicted labels is more important than the morphological information of the predictions in the early stage of the domain adaptation, and vice versa in the later stage.

The network parameters of the MT target model θ' at training step n as the exponential moving average of successive weights θ of the student target model, as in Eq. (7).

$$\theta'_n = \gamma \theta'_{n-1} + (1 - \gamma)\theta_n, \tag{7}$$

where γ is a smoothing coefficient hyperparameter.

3   Experiments

3.1   Datasets

To evaluate the versatility and robustness of the methods as domain adaptation, we prepared two datasets in a wide variety of source-target domain combinations. We have image (2-dimensional) and volume (3-dimensional) datasets, but we use 2-dimensional deep neural network architectures for all datasets. In the case of segmentation of 3-dimensional datasets, we perform the segmentation on a slice-by-slice basis. All tasks are binary segmentation, i.e., foreground and background classes. Examples of the images in the datasets are shown in Fig. 2. The number of samples in the source and target regions for each dataset is summarized in Table 1.

MRI Brain. This dataset was created from the BraTS2018 dataset [1,3,14]. The BraTS2018 dataset contains routine clinically acquired 3T multimodal brain MRI scans and ground truth labels annotated by expert neuroradiologists. The BraTS2018 dataset includes cases for two disease grades, four modalities, and three classes of annotation. We selected low-grade glioma (LGG) cases and two modalities (T2-weighted (T2) and T2 fluid attenuated inversion recovery (FLAIR)). We selected "whole tumor" (WT) as the foreground mask. The WT describes the complete extent of the disease. Each volume was normalized to zero mean and unit variance as a preprocessing step. The FLAIR images and segmentation masks were used as source domain samples and the T2 images as target domain samples. This dataset can be used for domain adaptation between different modalities, i.e., T2 and Flair in MRI.

Fundus. This dataset was created from the Retinal Fundus Glaucoma Challenge (REFUGE) dataset [2,4,8]. The REFUGE dataset consists of retinal fundus images and segmentation masks for the optic cup and optic disc regions. The dataset includes 400 training samples of 2124×2056 pixels acquired with a Zeiss Visucam 500 camera and 400 validation samples of 1634×1634 pixels acquired with a Canon CR-2 camera. We used the samples from the Canon camera as the source domain and the samples from the Zeiss camera as the target domain. As in a similar procedure in [11], we cropped a region in each image such that the center of the optic disc was the center of the cropped area and resized the cropped region to 384×384 pixels. The size of the cropped region was 600×600 for the Zeiss camera and 500×500 for the Canon camera. We used the optic cup segmentation mask as the foreground. The images were normalized using 128 as the mean and 64 as the standard deviation. This dataset can be used for domain matching between different manufacturers in the same modality, i.e. retinal fundus camera.

3.2   Experimental Conditions

We use U-Net [17] as the basic segmentation network and replace its encoder part as EfficientNet-B4 [18]. It is used for the models of the source and target

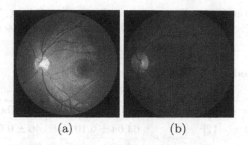

(a)   (b)

Fig. 2. Examples of images in datasets. (a) MRI Brain. (b) Fundus. Left and right sides are samples from source and target domains, respectively, in each dataset.

Table 1. Number of samples in datasets. The numbers in MRI Brain and MRI Prostate show the number of 3D volumes.

Dataset	Source domain	Target domain
MRI Brain	75	75
Fundus	400	400

domains. Note that the different architectures can be used for source and target domains in the BBUDA setting.

We trained source domain models using source domain datasets. A combination of cross-entropy loss and dice loss is used as the loss function in the source model training. Eighty percent of the source domain samples are used as training data, and the remaining samples are used as validation data for each dataset. We used the model with the lowest loss value for the validation data in the domain adaptation phase. The initial learning rate was 3×10^{-4} and was determined by grid search. As for the other experimental conditions, the number of epochs was 100, the optimizer was AdamW [13], and the learning rate was varied with cosine annealing.

In the domain adaptation phase, the number of epochs was 20, and the learning rate was multiplied by 0.1 every 10 epochs. The initial learning rate was 1×10^{-4}. These hyperparameters were determined by grid search. The optimizer was AdamW.

We compared our proposed method with the method proposed in [12] (referred to as "Liu et al.") as the state of the art for the segmentation task in BBUDA. We also evaluated the performance of the source domain model (referred to as "source only") as a lower bound and with supervised learning using ground truth for the target domain (referred to as "oracle") as an upper bound. As ablation studies, we evaluated the performance of our proposed method without some components, i.e., MT and Dice loss.

The evaluation was done by cross-validating the dataset in the target domain. Each dataset was divided into five groups. We used four groups for training and one group for testing. Thus, we had five folds for cross-validation and the performance was evaluated with the mean and standard deviation values of

Table 2. Experimental results. The metric is the Dice coefficient. The numbers mean "mean ± standard deviation". MT and DL stand for Mean Teacher and Dice Loss, respectively.

Method	MT	DL	Dataset	
			MRI Brain	Fundus
Source only	–	–	57.02 ± 0.043	85.52 ± 0.016
Liu et al. [12]	–	–	64.64 ± 0.102	86.46 ± 0.008
Ours	✓	–	64.68 ± 0.097	86.76 ± 0.007
	✓	✓	**66.02** ± 0.086	**87.26** ± 0.005
Oracle	–	–	82.82 ± 0.032	89.28 ± 0.005

the test data in five folds. We used dice coefficients as the evaluation metric for the test data. We implemented our method using PyTorch v1.12.1 [15] and Lightning v1.7.7 including conventional methods for comparison. We used an Nvidia RTX3090 GPU (24 GB memory).

3.3 Results and Discussion

Table 2 summarizes the experimental results. As shown in Table 2, our proposed method outperformed the state-of-the-art method (Liu et al. [12]) for all datasets.

For two datasets, "ours (MT+DL)" showed the best performance, and the improvement over Liu et al. was about 1.4 points in the mean Dice coefficient. When the domain gap between the source and target domains was small, i.e., the difference between the Dice coefficients of "source only" and "oracle" was small, such as the Fundus dataset, the segmentation performance of our proposed method was around the middle of "source only" and "oracle". On the other hand, when the domain gap between the source and target domains was large, i.e., the difference between the Dice coefficients of "source only" and "oracle" was large, such as the MRI Brain dataset, the segmentation performance of our proposed method was below the middle of 'source only" and "oracle". This is a limitation of our method.

Figure 3 shows examples of segmentation results for the target domain images. It can be seen that our proposed method improves the segmentation performance com-pared to Liu et al. For the MRI Brain dataset, the protruding shape in the upper left corner of the ground truth label is not detected by Liu et al. while it is detected by our proposed method (MT+DL). For the Fundus dataset, the concave shape in the lower right corner of the ground truth label is not detected by Liu et al. while it is detected by our proposed method (MT and MT+DL).

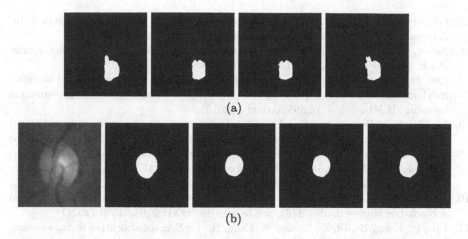

(a)

(b)

Fig. 3. Examples of segmentation results for the target domain images. (a) MRI Brain da-taset. (b) Fundus dataset. The images show the input target domain image, ground truth, prediction by Liu et al., prediction by ours (MT), prediction by ours (MT+DL) from left to right in each row.

4 Conclusions

In this paper, we proposed a method to improve the performance of the black-box unsupervised domain adaptation method for medical image segmentation. We introduced a mean teacher algorithm to stabilize the training during the training of the target domain model, and also introduced a dice loss to account for morphological information during the training of the target domain model. We conducted experiments on two datasets containing different types of source-target domain combinations and confirmed that our method outperformed the state-of-the-art in all datasets.

In the future, our method should be evaluated using more types of source-target domain combinations, datasets with multiple labels, and datasets with 3D volumes.

Acknowledgements. This work was supported by JSPS KAKENHI Grant Number JP22K12237.

References

1. Multimodal Brain Tumor Segmentation Challenge 2018. https://www.med.upenn.edu/sbia/brats2018/data.html
2. REFUGE: RETINAL FUNDUS GLAUCOMA CHALLENGE. https://ieee-dataport.org/documents/refuge-retinal-fundus-glaucoma-challenge
3. Bakas, S., et al.: Advancing the cancer genome atlas glioma MRI collections with expert segmentation labels and radiomic features. Sci. Data 4(1), 1–13 (2017)
4. Diaz-Pinto, A., et al.: Refuge challenge: a unified framework for evaluating automated methods for glaucoma assessment from fundus photographs (2020)

5. Grandvalet, Y., Bengio, Y.: Semi-supervised learning by entropy minimization. In: Advances in Neural Information Processing Systems, vol. 17 (2004)
6. Kouw, W.M., Loog, M.: A review of domain adaptation without target labels. IEEE Trans. Pattern Anal. Mach. Intell. **43**(3), 766–785 (2019)
7. Lee, D.H., et al.: Pseudo-label: The simple and efficient semi-supervised learning method for deep neural networks. In: Workshop on challenges in representation learning, ICML. vol. 3, p. 896. Atlanta (2013)
8. Li, F., et al.: Development and clinical deployment of a smartphone-based visual field deep learning system for glaucoma detection. NPJ Dig. Med. **3**(1), 123 (2020)
9. Liang, J., Hu, D., Feng, J., He, R.: Dine: domain adaptation from single and multiple black-box predictors. In: Proceedings of the IEEE/CVF Conference on Computer Vision and Pattern Recognition, pp. 8003–8013 (2022)
10. Liang, J., Hu, D., He, R., Feng, J.: Distill and fine-tune: effective adaptation from a black-box source model. **1**(3), arXiv preprint arXiv:2104.01539 (2021)
11. Liu, P., Kong, B., Li, Z., Zhang, S., Fang, R.: CFEA: collaborative feature ensembling adaptation for domain adaptation in unsupervised optic disc and cup segmentation. In: Shen, D., et al. (eds.) MICCAI 2019. LNCS, vol. 11768, pp. 521–529. Springer, Cham (2019). https://doi.org/10.1007/978-3-030-32254-0_58
12. Liu, X., et al.: Unsupervised black-box model domain adaptation for brain tumor segmentation. Front. Neurosci. **16**, 837646 (2022)
13. Loshchilov, I., Hutter, F.: Decoupled weight decay regularization. arXiv preprint arXiv:1711.05101 (2017)
14. Menze, B.H., et al.: The multimodal brain tumor image segmentation benchmark (brats). IEEE Trans. Med. Imaging **34**(10), 1993–2024 (2014)
15. Paszke, A., et al.: Pytorch: an imperative style, high-performance deep learning library. In: Advances in Neural Information Processing Systems, vol. 32 (2019)
16. Peng, Q., Ding, Z., Lyu, L., Sun, L., Chen, C.: Toward better target representation for source-free and black-box domain adaptation. arXiv preprint arXiv:2208.10531 (2022)
17. Ronneberger, O., Fischer, P., Brox, T.: U-net: convolutional networks for biomedical image segmentation. In: Navab, N., Hornegger, J., Wells, W.M., Frangi, A.F. (eds.) MICCAI 2015, Part III. LNCS, vol. 9351, pp. 234–241. Springer, Cham (2015). https://doi.org/10.1007/978-3-319-24574-4_28
18. Tan, M., Le, Q.: Efficientnet: rethinking model scaling for convolutional neural networks. In: International Conference on Machine Learning, pp. 6105–6114. PMLR (2019)
19. Tarvainen, A., Valpola, H.: Mean teachers are better role models: weight-averaged consistency targets improve semi-supervised deep learning results. In: Advances in Neural Information Processing Systems, vol. 30 (2017)
20. Toldo, M., Maracani, A., Michieli, U., Zanuttigh, P.: Unsupervised domain adaptation in semantic segmentation: a review. Technologies **8**(2), 35 (2020)
21. Wang, R., Lei, T., Cui, R., Zhang, B., Meng, H., Nandi, A.K.: Medical image segmentation using deep learning: a survey. IET Image Proc. **16**(5), 1243–1267 (2022)
22. Xiao, C., Li, B., Zhu, J.Y., He, W., Liu, M., Song, D.: Generating adversarial examples with adversarial networks. arXiv preprint arXiv:1801.02610 (2018)
23. Zhang, H., Zhang, Y., Jia, K., Zhang, L.: Unsupervised domain adaptation of black-box source models. arXiv preprint arXiv:2101.02839 (2021)

PLST: A Pseudo-labels with a Smooth Transition Strategy for Medical Site Adaptation

Tomer Bar Natan[1], Hayit Greenspan[1,2], and Jacob Goldberger[3(✉)]

[1] Tel-Aviv University, Tel-Aviv, Israel
tomerb5@mail.tau.ac.il, hayitg@gmail.com
[2] Tel Aviv University, Israel and Mount Sinai, New York, USA
[3] Bar-Ilan University, Ramat-Gan, Israel
jacob.goldberger@biu.ac.il

Abstract. This study addresses the challenge of medical image segmentation when transferring a pre-trained model from one medical site to another without access to pre-existing labels. The method involves utilizing a self-training approach by generating pseudo-labels of the target domain data. To do so, a strategy that is based on a smooth transition between domains is implemented where we initially feed easy examples to the network and gradually increase the difficulty of the examples. To identify the level of difficulty, we use a binary classifier trained to distinguish between the two domains by considering that target images easier if they are classified as source examples. We demonstrate the improved performance of our method on a range of medical MRI image segmentation tasks. When integrating our approach as a post-processing step in several standard Unsupervised Domain Adaptation (UDA) algorithms, we consistently observed significant improvements in the segmentation results on test images from the target site.

Keywords: domain shift · domain adaptation · self-training · pseudo-labels

1 Introduction

The application of deep learning systems to real-world problems is hindered by the drop in performance when a network trained on data from one domain is applied to data from a different domain. This is known as the domain shift problem. There are two main types of domain shift namely covariate shift [20] and label shift [13]. In this study, we focus on covariate shift, a scenario where the distribution of features changes across domains, but the distribution of labels remains constant given the features. The issue of domain shift is particularly critical in medical imaging where the accuracy of a model trained on data from one medical facility decreases when applied to data from a different site [5]. Each instance of data collection, such as an MRI machine, can be considered its own domain due to variations in the MRI machine between different vendors and variations in the scanning process in different medical sites. Collecting enough annotated data for each new site is an impractical solution due to cost and specialization requirements. This problem currently hinders the widespread automation of medical imaging techniques.

In an Unsupervised Domain Adaptation (UDA) setup we have access to data from the target domain but without labels. There are various techniques for adapting the

L. Koch et al. (Eds.): DART 2023, LNCS 14293, pp. 31–40, 2024.
https://doi.org/10.1007/978-3-031-45857-6_4

segmentation model in UDA, such as using adversarial learning to acquire domain-invariant features for image-to-image translation techniques such as CycleGAN [11], modulating the statistics in the Batch-Normalization layer [8] or layer-wise activation matching between domains [7]. Updated reviews on UDA for segmentation of scenery and medical images can be found in [3] and [5].

Self-training [9] converts model predictions into pseudo labels. These pseudo labeled images are then combined in semi-supervised setups with labeled images to train an improved model. Self-training was also found to be a promising method for UDA of segmentation of scenery images [6,24] and medical images [2]. Several studies have applied pseudo labels to cross modality UDA (e.g. adapting a segmentation system of T1 MRI images to T2 MRI images) [22]. In a recent UDA challenge on cross modality medical imaging segmentation, the best-performing team found that applying a pseudo-labels post-processing step led to a significant performance improvement [16].

In this study we propose a pseudo labeling self-learning procedure that can be combined with UDA methods as a post-processing step. Our domain adaptation method involves incorporating target domain examples into the self-learning process based on their similarity to the source domain. We applied our method to two medical MRI image segmentation tasks as a post-processing step after a variety of UDA algorithms, and found that this consistently and significantly improved each UDA segmentation results on test images from the target site.[1]

2 Pseudo-labels with a Smooth Transition

The concept of self-training, also known as pseudo-labeling, was initially developed in the context of semi-supervised learning. This involves training a network on a small set of labeled data, and then using the network to generate pseudo labels for the remaining unlabeled data. The labeled and pseudo-labeled data are then used together to iteratively train the network, where improved pseudo labels are generated at each iteration.

In semi-supervised learning, both the labeled and unlabeled data are sampled from the same distribution and are equally relevant to the learning process. However, in unsupervised domain adaptation (UDA), there is a crucial difference between the labeled and unlabeled data. The labeled data comes from the source domain and may not accurately represent the target domain. Despite this shortcoming, current applications of pseudo-labels to UDA use all instances from the source and target domains [6,24–26]. Self-training produces pseudo-labels that are noisy. Most recent progress in pseudo-labeling is mostly focused on attempts to solve the label noise problem by selecting target data based on pseudo-label confidence. In the case of classification, the most confident images are selected and in the case of segmentation, the most confident pixels within each image are selected, see e.g. [12] [17].

Domain adaptation techniques modify a network's parameters to match the target domain by following a continuous trajectory in the parameter space. Our proposed PL

---

[1] Code for our implementation including data pre processing and trained models is available at: https://github.com/TomerBarNatan/PLST.

Algorithm 1. Pseudo-Labels with a Smooth Transition Algorithm (PLST)

Input: labeled data for the source domain and unlabeled data from the target domain, and a
    parameter-set θ fine-tuned by a UDA algorithm.
- Train a binary classifier to distinguish between source and target domains.
- Sort the data of each domain according to the probability to be classified as a target from least
    to most.
for k in 1 until convergence **do**
    - Generate pseudo labels: $\hat{y}_{k,1}, ..., \hat{y}_{k,r_k}$ using the current parameter-set θ.
    - Fine-tune θ by minimizing the following loss function:

$$L_k(\theta) = \alpha_k \cdot \frac{1}{r_k} \sum_{j=1}^{r_k} L(x_j^t, \hat{y}_{kj}^t, \theta) + (1-\alpha_k) \cdot \frac{1}{n-l} \sum_{i=l}^{n} L(x_i^s, y_i^s, \theta)$$

end for

---

scheme chooses samples that closely correspond to the current position on this trajectory, taking into account their appearance as either source or target samples. To identify suitable samples, we train a binary classifier that distinguishes between source and
target, utilizing the soft classification probabilities to determine whether each sample
should be treated as a target instance. The aim of our approach is twofold: first, to select
samples that align well with the current network, thus reducing prediction noise, and
second, to progressively shift the model's focus from the source domain towards the
target domain.

We begin by utilizing an image segmentation model, which can be either the original
model trained on source domain data or a model obtained through one of the numerous
available UDA algorithms. The first step is to train a binary classifier that can distinguish between samples from the source and target domains. The domain classifier is
focused on the segmentation task; hence the features are taken from the segmentation
U-net bottleneck layer. We sort the source domain samples according to their probability of being classified as a target from the least to the most, denoted as $x_1^s, ..., x_n^s$.
Our approach involves selecting the subset of images from the source domain with the
highest probability of being classified as target examples. These labeled images serve
as the training data to adapt the network to the target domain.

The domain classifier is also used to sort the unlabeled target domain data from
the most similar to the source domain to the least similar. We denote the sorted set as
$x_1^t, ..., x_m^t$. While all the target examples are equally relevant, those that are closer to
the source domain are easier when tuning a network that was originally trained on the
source domain. Thus, they are more suitable for the network in its intermediate phase
from the source domain to the target. In this approach, we begin with target samples that
are more easy and at each iteration of the pseudo-label scheme, as the network becomes
more adapted to the target, we add more challenging examples in the sense that they
look different from the source examples.

Each iteration of the pseudo-labels scheme consists of two steps. In the first step
we apply the current model on easy target data $x_1^t, ..., x_{r_k}^t$ to obtain pseudo labels
$\hat{y}_1^t, ..., \hat{y}_{r_k}^t$, where r_k is a monotonically increasing function of the iteration index k.

In the second step we fine-tune the network using both the relevant source data and the easy examples from the target with their current pseudo labels. The fine-tuning loss at the k-th iteration is:

$$L_k(\theta) = \alpha_k \cdot \frac{1}{r_k} \sum_{j=1}^{r_k} L(x_j^t, \hat{y}_{kj}^t, \theta) + (1 - \alpha_k) \cdot \frac{1}{n-l} \sum_{i=l}^{n} L(x_i^s, y_i^s, \theta) \qquad (1)$$

where a_k is a monotonically increasing function of k which ensures a smooth transition from the source domain to the target. $L(x, y, \theta)$ is the pixelwise cross-entropy of the model's parameters θ. The hyper-parameter l remains constant throughout the process and determines the number of source examples to select based on the ordering of the target-similarity. In practice, we found that this method is not highly sensitive to the exact values of a_k and r_k, whereas taking small l can significantly harm the results. In the next section we report the values that worked well in all the experiments.

The proposed self-training algorithm which we dub the Pseudo-Labels with a Smooth Transition Algorithm (PLST) is shown in Algorithm box 1.

3   Experimental Results

We applied our self-training method as a post-processing step to various UDA methods. We show below results on two different medical imaging segmentation tasks. In all our experiments we assumed that we had labeled data from the source site, a model trained on the source site data and unlabeled training data from the target site. The performance was evaluated on a labeled test dataset from the target site. In principle, the proposed method can be combined with any UDA method. We chose a representative UDA algorithm from the four most dominant approaches today that deal with UDA (image statistics, domain shift minimization in feature space and feature alignment adversarial networks).

- *AdaBN*: recalculating the batch normalization statistics on the target site [10].
- *Seg-JDOT*: aligning the two distributions of the source and the target sites by an optimal transport algorithm [1].
- *AdaptSegNet*: aligning feature space using adversarial learning [21].
- *AIVA*: aligning the feature space using clustering in the embedded space [4].

In addition, we report the results of our method when applied directly to the pre-trained model without any adaptation. We also trained a network on the target site using the labels from the training data of the target site, thereby setting an *upper bound* for both the UDA and our methods. Note that the purpose was not to compare various UDA algorithms, but rather to demonstrate that our procedure can be utilized with any UDA algorithm to achieve even greater performance enhancements.

Implementation Details: In all our experiments we pre-trained a standard U-net network on the data from the source domain using 5K steps and ensured that the models reached the loss-function plateau. We then applied a UDA algorithm followed by our method. For the domain classifier we trained a model consisting of three convolution layers followed by a fully connected layer. In all our experiments we took 10% of the source data that were classified as target with the highest probability. The initial value of r_0 was 10% of the total number of target data and then was linearly increased to 100%. We initiated α_0 with 0.5 and was linearly increased to 1. In addition, we updated the target pseudo labels 45 times and trained the network for 30 epochs after each update. These numbers were verified empirically since we saw no significant improvement in the loss by increasing them, in other words, the algorithm indeed reached the loss plateau and converged. The PLST has a small runtime complexity relative to the UDA algorithms and it took approximately 10% of the entire adaptation learning process.

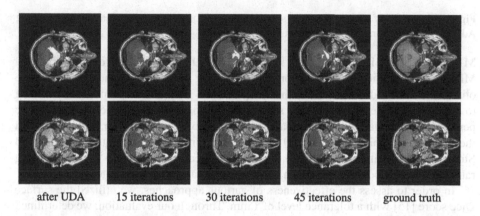

| after UDA | 15 iterations | 30 iterations | 45 iterations | ground truth |

Fig. 1. Two examples of MRI skull striping segmentation result of images from the target domain after PLST iterations.

Table 1. Segmentation surface-Dice results on the brain MRI dataset CC359 [19]. Results are shown for each UDA method with (+PLST) and without pseudo-labels post processing. Each line shows the average results on five source sites.

target site	target site model	source site model		unsupervised adaptation							
			+PLST	AdaBN	+PLST	Seg-JDOT	+PLST	AdaptSegNet	+PLST	AIVA	+PLST
Siemens, 1.5T	80.13	58.44	62.76	62.20	68.11	64.27	69.01	63.41	73.55	67.43	**75.52**
Siemens, 3T	80.19	59.65	63.94	58.82	64.21	61.83	66.31	57.57	70.19	66.31	**72.44**
GE, 1.5T	81.58	38.96	49.21	58.04	63.94	50.06	58.28	57.71	68.24	59.95	**69.36**
GE, 3T	84.16	56.27	58.11	54.23	60.80	59.82	66.15	55.94	70.23	65.04	**71.92**
Philips, 1.5T	84.00	56.38	60.09	73.01	74.41	69.59	75.86	68.22	75.62	75.68	**80.48**
Philips, 3T	82.19	41.93	44.48	51.18	58.79	50.30	59.46	50.51	59.54	54.34	**63.50**
Average	82.04	51.94	56.43	59.58	65.04	59.31	65.84	58.89	69.56	64.79	**72.20**

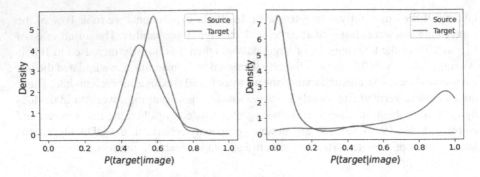

Fig. 2. Histograms of domain classifier soft decisions of source and target images after applying AdaptSegNet (left) and AdaBN (right) UDA algorithms.

MRI Skull Stripping: The dataset known as CC359 [19] includes a collection of 359 MR images specifically focusing on skull stripping of the head. These images were obtained from six different sites, which experienced a significant domain shift leading to a notable decline in the accuracy scores [18]. As part of the preprocessing stage, we performed interpolation to achieve a voxel spacing of $1 \times 1 \times 1$ mm, and then scaled the intensities within the range of 0 to 1. As an optimization technique, we employed Stochastic Gradient Descent (SGD) with an initial learning rate of 10^{-6}. This learning rate was decayed polynomially with a power of 0.9, and we used a batch size of 16.

In order to assess the effectiveness of various approaches, we utilized the surface Dice score [15] with a tolerance level of 1 mm. Through our evaluation, we determined that the surface Dice score proved to be a more appropriate metric for measuring the success of brain segmentation compared to the conventional Dice score, as demonstrated in previous work [23].

We partitioned each target site into a training set and a test set. Employing 30 pairs of source and target sites, we computed the average results for each target site. The surface-Dice scores obtained from the test portion of the target domain are presented in Table 1. It shows that for all the UDA algorithms, our PLST significantly improved performance. This served to confirm that our self-training strategy is indeed complementary to current domain adaptation techniques for medical image segmentation. Two examples of skull stripping segmentation are shown in Fig. 1.

The assessment of improvement percentages in various UDA algorithms based on Table 1 reveals a disparity in the different fine-tuning strategies. For instance, in AdaptSetNet there was an average improvement of 18.11% but only 9.16% for adaBN. Figure 2 lists the typical densities of soft decision outcomes for the target domain of the classifier, as produced by both algorithms. It shows that the feature alignment algorithm had a lower level of certainty in its classification results than the other strategy, where the classifier could easily distinguish between source and target features. This finding suggests that a network trained with a feature alignment strategy, extracts image features in a way that improves the classifier's reliability in terms of identifying similarities between different domains. PLST should thus generate better results for UDA algorithms that aim to minimize the gap between the source and target features.

Prostate MRI Segmentation. In order to validate the robustness of the PLST method, we conducted an evaluation using a multi-source single-target configuration. For the

Table 2. Segmentation Dice results on the prostate MRI dataset [14]. Results are shown for each UDA method with (+PLST) and without pseudo-labels post processing.

target site	target site model	source site model		unsupervised adaptation					
			+PLST	AdaBN	+PLST	AdaptSegNet	+PLST	AIVA	+PLST
Site A	87.48	80.63	**82.75**	79.03	82.44	76.63	81.28	79.94	82.02
Site B	85.62	59.84	64.20	73.32	79.29	64.45	73.84	78.50	**82.91**
Site C	78.86	58.84	63.03	68.08	71.85	67.64	73.10	69.67	**72.46**
Site D	85.30	61.78	65.98	66.99	73.59	77.39	80.96	78.88	**81.90**
Site E	80.55	77.72	79.44	79.24	**80.53**	79.13	80.27	79.65	80.42
Site F	86.10	78.41	80.60	53.36	61.16	80.88	84.04	82.72	**85.79**
Average	83.98	69.80	72.66	70.21	74.81	74.35	78.91	78.67	**80.91**

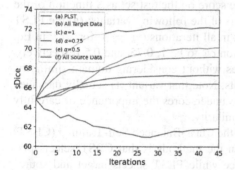 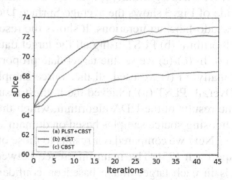

Fig. 3. (left) sDice results of different variants of PLST as a function of the number of PL updating. (right) Comparison between PLST and CBST.

purpose of prostate MRI segmentation, we utilized a publicly available multi-site dataset consisting of prostate T2-weighted MRI data, accompanied by segmentation masks. This dataset was collected from various data sources, resulting in a noticeable distribution shift. Additional information regarding the data and imaging protocols from the six distinct sites can be found in [14]. Samples of sites A,B were from the NCI-ISBI13 dataset, samples of site C were taken from the I2CVB dataset, and samples of sites D,E,F were from the PROMISE12 dataset.

As part of the pre-processing steps, we standardized each sample by normalizing its intensity values to have a zero mean and a unit variance before feeding it into the network. For every target site, we utilized the remaining five sites as the source data. The results were then computed for all six potential target sites. As the optimization method, we employed ADAM with an initial learning rate of 10^{-6}, a weight decay of 0.9, and a batch size of 8. In order to assess the effectiveness of the different approaches, we utilized the Dice Score as the evaluation metric.

The dataset was divided into training and test sets for each individual site. Table 2 shows the comparative performances of each target site. We could not achieve convergence for seg-JDOT [1] on this particular dataset, potentially due to insufficient data. Hence, we excluded seg-JDOT from the reported results.

In some sites, there was no big difference between the target and the source model results. Consequently, this led to relatively weaker results for some of the UDA methods in those cases. Table 2 shows that for all UDA algorithms, our PLST procedure significantly improved the performance of the original UDA method and it is closing the gap from supervised training on the target domain.

Ablation Study. We investigate the contribution of each element in the PLST algorithm and assess its impact on the overall performance. Our algorithm has three key components: it moves gradually from easy to difficult target samples, enforces a monotonic increase of the target sample proportion in the loss function and it makes a careful selection of the relevant source examples. To validate the contribution of each component we ran the PLST without them individually. The analysis applied PLST on the AIVA UDA algorithm [4] using the 30 source-target pairs of the CC359 dataset. The left side of Fig. 3 shows the average Surface Dice score on the test set as a function of the pseudo-labeling iterations. It shows the results of the following variants: (a) the PLST algorithm, (b) PLST using all the target data in all iterations ($r_k = m$ for all k in Eq. (1)). In (c)-(e) we set the target data proportion α_k to be 1, 0.75, and 0.5 respectively. Finally, in (f) we used all the source samples without considering domain similarity. Overall, PLST (a) achieved the highest results. Note that variant (f) actually damaged the results of the UDA algorithm, which thus underscores the importance of carefully choosing source samples based on domain similarity.

Next we compared our results to those of the Class-Balanced Self-Training (CBST) approach [25]. The main difference between the methods is that CBST selects pixels in each target image based on confidence while PLST selects target and source images based on their similarity to the other domain. We ran CBST for 45 iterations after running AIVA as UDA algorithm using the 30 source-target pairs of the CC359 dataset. We gradually increased the proportion of selected pixels from 20% to 50% as the paper suggests. Figure 3 on the right shows the average Surface Dice score on the test set as a function of iterations. Experiment (a) involved the fusion of PLST and CBST approaches by selecting examples based on domain similarity and identifying the portions of pixels with the highest confidence. The output of the PLST and CBST algorithms is shown in (b) and (c), respectively. Our experimentation demonstrates that both methods enhance the results, but the PLST approach has a more substantial impact. The superior performance of the hybrid approach implies that it is crucial to consider both domain similarity and pixel confidence for optimal results.

To conclude, we developed a pseudo-label technique specifically designed for domain adaptation. By leveraging both source and target samples, we gradually transitioned the network from the source domain to the target domain, resulting in improved performance on several segmentation tasks. Note that while our PLST adaptation method was evaluated in the context of medical imaging segmentation, it is a general technique that can be applied to other medical imaging scenarios, including classification and regression. Our work revealed an interesting observation about the dynamics of pseudo labels between iterations. Specifically, we found that the changes in pseudo labels were concentrated at the edges of segmentation areas, while the pixels in the middle remained stable. This pattern was consistent across all variations of our experiments. Leveraging this observation could lead to more effective optimization in edge pixels, which could be a promising research direction.

References

1. Ackaouy, A., et al.: Unsupervised domain adaptation with optimal transport in multi-site segmentation of multiple sclerosis lesions from MRI data. Frontiers Comput. Neurosci. **14**, 19 (2020)
2. Bar Natan, T., Greenspan, H., Goldberger, J.: PLPP: a pseudo labeling post-processing strategy for unsupervised domain adaptation. In: The IEEE International Symposium on Biomedical Imaging (ISBI) (2023)
3. Csurka, G., Volpi, R., Chidlovskii, B.: Unsupervised domain adaptation for semantic image segmentation: a comprehensive survey. arXiv preprint arXiv:2112.03241 (2021)
4. Goodman, S., Serlin, S.K., Greenspan, H., Goldberger, J.: Unsupervised site adaptation by intra-site variability alignment. In: Kamnitsas, K., et al. (eds.) DART 2022. LNCS, vol. 13542, pp. 56–65. Springer, Cham (2022). https://doi.org/10.1007/978-3-031-16852-9_6
5. Guan, H., Liu, M.: Domain adaptation for medical image analysis: a survey. IEEE Trans. Biomed. Eng. **69**(3), 1173–1185 (2021)
6. Hoyer, L., Dai, D., Van Gool, L.: Daformer: improving network architectures and training strategies for domain-adaptive semantic segmentation. In: Proceedings of the IEEE Conference on Computer Vision and Pattern Recognition (CVPR) (2022)
7. Huang, H., Huang, Q., Krähenbühl, P.: Domain transfer through deep activation matching. In: Ferrari, V., Hebert, M., Sminchisescu, C., Weiss, Y. (eds.) ECCV 2018. LNCS, vol. 11220, pp. 611–626. Springer, Cham (2018). https://doi.org/10.1007/978-3-030-01270-0_36
8. Kasten-Serlin, S., Goldberger, J., Greenspan, H.: Adaptation of a multisite network to a new clinical site via batch-normalization similarity. In: The IEEE International Symposium on Biomedical Imaging (ISBI) (2022)
9. Lee, D.H., et al.: Pseudo-label: the simple and efficient semi-supervised learning method for deep neural networks. In: Workshop on challenges in representation learning, ICML (2013)
10. Li, Y., Wang, N., Shi, J., Hou, X., Liu, J.: Adaptive batch normalization for practical domain adaptation. Pattern Recogn. **80**, 109–117 (2018)
11. Li, Y., Yuan, L., Vasconcelos, N.: Bidirectional learning for domain adaptation of semantic segmentation. In: Proceedings of the IEEE Conference on Computer Vision and Pattern Recognition (CVPR) (2019)
12. Liang, J., Hu, D., Feng, J.: Domain adaptation with auxiliary target domain-oriented classifier. In: Proceedings of the IEEE Conference on Computer Vision and Pattern Recognition (CVPR) (2021)
13. Lipton, Z., Wang, Y.X., Smola, A.: Detecting and correcting for label shift with black box predictors. In: International Conference on Machine Learning (ICML) (2018)
14. Liu, Q., Dou, Q., Heng, P.-A.: Shape-aware meta-learning for generalizing prostate MRI segmentation to unseen domains. In: Martel, A.L., et al. (eds.) MICCAI 2020. LNCS, vol. 12262, pp. 475–485. Springer, Cham (2020). https://doi.org/10.1007/978-3-030-59713-9_46
15. Nikolov, S., et al.: Deep learning to achieve clinically applicable segmentation of head and neck anatomy for radiotherapy. CoRR abs/1809.04430 (2018)
16. Shin, H., Kim, H., Kim, S., Jun, Y., Eo, T., Hwang, D.: Cosmos: cross-modality unsupervised domain adaptation for 3d medical image segmentation based on target-aware domain translation and iterative self-training. arXiv preprint arXiv:2203.16557 (2022)

17. Shin, I., Woo, S., Pan, F., Kweon, I.S.: Two-phase pseudo label densification for self-training based domain adaptation. In: Vedaldi, A., Bischof, H., Brox, T., Frahm, J.-M. (eds.) ECCV 2020. LNCS, vol. 12358, pp. 532–548. Springer, Cham (2020). https://doi.org/10.1007/978-3-030-58601-0_32

18. Shirokikh, B., Zakazov, I., Chernyavskiy, A., Fedulova, I., Belyaev, M.: First U-net layers contain more domain specific information than the last ones. In: Albarqouni, S., et al. (eds.) DART/DCL -2020. LNCS, vol. 12444, pp. 117–126. Springer, Cham (2020). https://doi.org/10.1007/978-3-030-60548-3_12

19. Souza, R., et al.: An open, multi-vendor, multi-field-strength brain MR dataset and analysis of publicly available skull stripping methods agreement. Neuroimage **170**, 482–494 (2018)

20. Sugiyama, M., Nakajima, S., Kashima, H., Buenau, P., Kawanabe, M.: Direct importance estimation with model selection and its application to covariate shift adaptation. In: Advances in Neural Information Processing Systems (NeurIPs) (2007)

21. Tsai, Y.H., Hung, W.C., Schulter, S., Sohn, K., Yang, M.H., Chandraker, M.: Learning to adapt structured output space for semantic segmentation. In: Proceedings of the IEEE Conference on Computer Vision and Pattern Recognition (CVPR) (2018)

22. Xie, Q., et al.: Unsupervised domain adaptation for medical image segmentation by disentanglement learning and self-training. IEEE Trans. Med. Imaging (2022)

23. Zakazov, I., Shirokikh, B., Chernyavskiy, A., Belyaev, M.: Anatomy of domain shift impact on U-net layers in MRI segmentation. In: de Bruijne, M., et al. (eds.) MICCAI 2021. LNCS, vol. 12903, pp. 211–220. Springer, Cham (2021). https://doi.org/10.1007/978-3-030-87199-4_20

24. Zhang, P., Zhang, B., Zhang, T., Chen, D., Wang, Y., Wen, F.: Prototypical pseudo label denoising and target structure learning for domain adaptive semantic segmentation. In: Proceedings of the IEEE Conference on Computer Vision and Pattern Recognition (CVPR) (2021)

25. Zou, Y., Yu, Z., Vijaya Kumar, B.V.K., Wang, J.: Unsupervised domain adaptation for semantic segmentation via class-balanced self-training. In: Ferrari, V., Hebert, M., Sminchisescu, C., Weiss, Y. (eds.) ECCV 2018. LNCS, vol. 11207, pp. 297–313. Springer, Cham (2018). https://doi.org/10.1007/978-3-030-01219-9_18

26. Zou, Y., Yu, Z., Liu, X., Kumar, B.V., Wang, J.: Confidence regularized self-training. In: Proceedings of the IEEE International Conference on Computer Vision (ICCV) (2019)

Compositional Representation Learning for Brain Tumour Segmentation

Xiao Liu[1,2(✉)], Antanas Kascenas[1], Hannah Watson[1], Sotirios A. Tsaftaris[1,2,3],
and Alison Q. O'Neil[1,2]

[1] Canon Medical Research Europe Ltd., Edinburgh, UK
xiao.liu@mre.medical.canon
[2] School of Engineering, University of Edinburgh, Edinburgh EH9 3FB, UK
[3] The Alan Turing Institute, London, UK

Abstract. For brain tumour segmentation, deep learning models can
achieve human expert-level performance given a large amount of data
and pixel-level annotations. However, the expensive exercise of obtaining pixel-level annotations for large amounts of data is not always fea-
sible, and performance is often heavily reduced in a low-annotated data
regime. To tackle this challenge, we adapt a mixed supervision framework, vMFNet, to learn robust compositional representations using unsupervised learning and weak supervision alongside non-exhaustive pixel-
level pathology labels. In particular, we use the BraTS dataset to simu-
late a collection of 2-point expert pathology annotations indicating the
top and bottom slice of the tumour (or tumour sub-regions: peritumoural
edema, GD-enhancing tumour, and the necrotic/non-enhancing tumour)
in each MRI volume, from which weak image-level labels that indicate
the presence or absence of the tumour (or the tumour sub-regions) in
the image are constructed. Then, vMFNet models the encoded image
features with von-Mises-Fisher (vMF) distributions, via learnable and
compositional vMF kernels which capture information about structures
in the images. We show that good tumour segmentation performance
can be achieved with a large amount of weakly labelled data but only
a small amount of fully-annotated data. Interestingly, emergent learning
of anatomical structures occurs in the compositional representation even
given only supervision relating to pathology (tumour).

Keywords: Compositionality · Representation learning ·
Semi-supervised · Weakly-supervised · Brain tumour segmentation

1 Introduction

When a large amount of labelled training data is available, deep learning techniques have demonstrated remarkable accuracy in medical image segmentation
[2]. However, performance drops significantly when insufficient pixel-level annotations are available [16,17,24]. By contrast, radiologists learn clinically relevant
visual features from "weak" image-level supervision of seeing many medical scans

L. Koch et al. (Eds.): DART 2023, LNCS 14293, pp. 41–51, 2024.
https://doi.org/10.1007/978-3-031-45857-6_5

[1]. When searching for anatomy or lesions of interest in new images, they look for characteristic configurations of these clinically relevant features (or components). A similar compositional learning process has been shown to improve deep learning model performance in many computer vision tasks [9,11,25] but has received limited attention in medical applications.

In this paper, we consider a limited annotation data regime where few pixel-level annotations are available for the task of brain tumour segmentation in brain MRI scans. Alongside this, we construct slice-level labels for each MRI volume indicating the presence or absence of the tumour. These labels can be constructed from 2-point expert pathology annotations indicating the top and bottom slices of the tumour, which are fast to collect. We consider that pathology annotations are not only better suited to the task (tumour segmentation) but also to the domain (brain MRI) than the originally proposed weak supervision with anatomy annotations [15]; annotating the top and bottom slices for anatomical brain structures such as white matter, grey matter and cerebrospinal fluid (CSF) would be relatively uninformative about the configurations of structures within the image due to their whole brain distributions.

For the learning paradigm, we investigate the utility of learning compositional representations in increasing the annotation efficiency of segmentation model training. Compositional frameworks encourage identification of the visible semantic components (e.g. anatomical structures) in an image, requiring less explicit supervision (labels). We follow [11,15,18] in modelling compositional representations of medical imaging structures with learnable von-Mises-Fisher (vMF) kernels. The vMF kernels are learned as the cluster centres of the feature vectors of the training images, and the vMF activations determine which kernel is activated at each position. On visualising kernel activations, it can be seen that they approximately correspond to human-recognisable structures in the image, lending interpretability to the model predictions. Our contributions are summarised as:

- We refine an existing mixed supervision compositional representation learning framework, vMFNet, for the task of brain tumour segmentation, changing the weak supervision task from anatomy presence/absence to more domain-suited pathology presence/absence and simplifying the architecture and training parameters according to the principle of parsimony (in particular reducing the number of compositional vMF kernels and removing an original training subtask of image reconstruction).
- We perform extensive experiments on the BraTS 2021 challenge dataset [4,5,19] with different percentages of labelled data, showing superior performance of the proposed method compared to several strong baselines, both quantitatively (better segmentation performance) and qualitatively (better compositional representations).
- We compare weak pathology supervision with *tumour* labels to richer tumour *sub-region* labels, showing that the latter increases model accuracy for the task of tumour sub-region segmentation but also reduces the generality of

the compositional representation, which loses anatomical detail and increases in pathology detail, becoming more focused on the supervision task.

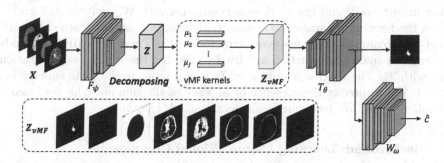

Fig. 1. Illustration of the brain tumour segmentation task using vMFBrain for compositional representation learning. We extract the weak supervision pathology labels (*presence or absence of tumour*) from 2-point brain tumour annotations; interestingly, learning of anatomical structures somewhat emerges even without supervision. Notation is specified in Sect. 3.

2   Related Work

Compositionality is a fundamental concept in computer vision, where it refers to the ability to recognise complex objects or scenes by detecting and combining simpler components or features [13]. Leveraging this idea, compositional representation learning is an area of active research in computer vision [27]. Early approaches to compositional representation learning in computer vision include the bag-of-visual-words model [12] and part-based models [11]. Compositional representation learning has been applied to fine-grained recognition tasks in computer vision, such as recognising bird species [9,23]. In addition, compositionality has been incorporated for robust image classification [11,25] and recently for compositional image synthesis [3,14]. Among these works, Compositional Networks [11], originally designed for robust classification under object occlusion, are easier to extend to pixel-level tasks as they estimate spatial and interpretable vMF likelihoods. Previous work integrates vMF kernels [11] for object localisation [26] and recently for nuclei segmentation (with the bounding box as supervision) in a weakly supervised manner [28]. More recently, vMFNet [18] applies vMF kernels for cardiac image segmentation in the domain generalisation setting. Additionally, vMFNet integrated weak labels indicating the presence or absence of cardiac structures and this gave improved performance [15]. We use similar types of weak image-level annotations but apply the vMF kernels to pathology segmentation and supervise with weak labels indicating the presence or absence of pathological structures.

3   Method

We apply vMFNet [15,18], as shown in Fig. 1, a model consisting of three modules: the feature extractor $\boldsymbol{F}_\psi$, the task network $\boldsymbol{T}_\theta$ (for brain tumour segmentation in our case), and the weak supervision network $\boldsymbol{W}_\omega$, where ψ, θ and ω denote the network parameters. Compositional components are learned as vMF kernels by decomposing the features extracted by $\boldsymbol{F}_\psi$. Then, the vMF likelihoods that contain spatial information are used to predict the tumour segmentation mask with $\boldsymbol{T}_\theta$. The voxel-wise output of $\boldsymbol{T}_\theta$ is also input to the weak supervision network $\boldsymbol{W}_\omega$ to predict the presence or absence of the tumour. This framework is detailed below. We term our implementation *vMFBrain*.

3.1   Background: Learning Compositional Components

To learn compositional components, the image features $\mathbf{Z} \in \mathbb{R}^{H \times W \times D}$ are first extracted by $\boldsymbol{F}_\psi$. H and W are the spatial dimensions and D is the number of channels. The feature vector $\mathbf{z}_i \in \mathbb{R}^D$ is defined as the normalised vector (i.e. $\|\mathbf{z}_i\| = 1$) across channels at position i on the 2D lattice of the feature map. Then, the image features are modelled with J vMF distributions. Each distribution has a learnable mean that is defined as vMF kernel $\boldsymbol{\mu}_j \in \mathbb{R}^D$. To ensure computational tractability, a fixed variance σ is set for all distributions. The vMF likelihood for the jth distribution at each position i is calculated as:

$$p(\mathbf{z}_i|\boldsymbol{\mu}_j) = \frac{e^{\sigma_j \boldsymbol{\mu}_j^T \mathbf{z}_i}}{C}, \text{ s.t. } \|\boldsymbol{\mu}_j\| = 1, \tag{1}$$

where C is a constant. This gives the vMF likelihood vector $\mathbf{z}_{i,vMF} \in \mathbb{R}^J$, a component of $\mathbf{Z}_{vMF} \in \mathbb{R}^{H \times W \times J}$, which determines which kernel is activated at each position. To update the kernels during training, the clustering loss $\mathcal{L}_{clu}$ is defined in [11] as:

$$\mathcal{L}_{clu}(\boldsymbol{\mu}, \mathbf{Z}) = -(HW)^{-1} \sum_i \max_j \boldsymbol{\mu}_j^T \mathbf{z}_i, \tag{2}$$

where the kernel $\boldsymbol{\mu}_j$ which is maximally activated for each feature vector $\mathbf{z}_i$ is found, and the distance between the feature vectors and their corresponding kernels is minimised by updating the kernels. Overall, feature vectors in different images corresponding to the same anatomical or pathological structure will be clustered and activate the same kernels. Hence, the vMF likelihoods $\mathbf{Z}_{vMF}$ for the same anatomical or pathological features in different images will be aligned to follow the same distributions (with the same means).

3.2   vMFBrain for Brain Tumour Segmentation

Taking the vMF likelihoods as input, a follow-on segmentation task module $\boldsymbol{T}_\theta$, is trained to predict the tumour segmentation mask, i.e. $\hat{\mathbf{Y}} = \boldsymbol{T}_\theta(\mathbf{Z}_{vMF})$. Firstly, we use direct strong supervision from the available pixel-level annotations $\mathbf{Y}$. Secondly, we define the weak supervision label c as a scalar (or a vector $\mathbf{c}$) which indicates the presence or absence of the tumour (or the presence or absence of the tumour sub-regions) in the 2D image slice. We use the output of the segmentation module as the input for a weak supervision classifier $\mathbf{W}_\omega$ i.e. $\hat{c} = \mathbf{W}_\omega(\hat{\mathbf{Y}})$. We train the classifier using $L1$ distance i.e. $\mathcal{L}_{weak}(\hat{c}, c) = |\hat{c} - c|_1$.

Overall, the model contains trainable parameters ψ, θ, ω and the vMF kernel means $\boldsymbol{\mu}$. The model (including all the modules) is trained **end-to-end** with the following objective:

$$\underset{\psi, \theta, \omega, \mu}{\operatorname{argmin}} \mathcal{L}_{clu} + \lambda_{Dice}\mathcal{L}_{Dice}(\mathbf{Y}, \hat{\mathbf{Y}})(\boldsymbol{\mu}, \mathbf{Z}) + \lambda_{weak}\mathcal{L}_{weak}(\hat{c}, c), \qquad (3)$$

where $\mathcal{L}_{Dice}$ is Dice loss [7,20]. We set $\lambda_{Dice} = 1$ when the ground-truth mask $\mathbf{Y}$ is available, otherwise $\lambda_{Dice} = 0$. We set λ_{weak} as 0.5 for the whole tumour segmentation task and λ_{weak} as 0.1 for the tumour sub-region segmentation task (values determined empirically).

4   Experiments

4.1   Dataset

We evaluate on the task of brain tumour segmentation using data from the BraTS 2021 challenge [4,5,19]. This data comprises native (T1), post-contrast T1-weighted (T1Gd), T2-weighted (T2), and T2 Fluid Attenuated Inversion Recovery (FLAIR) modality volumes for 1,251 patients from a variety of institutions and scanners. We split the data into train, validation and test sets containing 938, 62 and 251 subjects. The data has already been co-registered, skull-stripped and interpolated to the same resolution, each volume having 155 2D slices. Labels are provided for tumour sub-regions: the peritumoural edema (ED), the GD-enhancing tumour (ET), and the necrotic and non-enhancing tumour (NE). We additionally downscale all images to 128×128.

4.2   Baselines

We compare to the baselines **UNet** [22], **SDNet** [6] and **vMFNet** [18]. **SDNet** [6] is a semi-supervised disentanglement model with anatomy and modality encoders to separately encode the anatomical structure information and the imaging characteristics. The anatomical features are used as the input to the segmentor for the task of segmentation; the model is also trained with unlabelled data on the task of reconstructing the image by recombining the anatomy and modality factors. We compare to **vMFNet** with the architecture and training loss as described in [18]; this setup does not use weak supervision and has an additional image reconstruction module which we found empirically not to help performance (which thus we omit from vMFBrain).

4.3   Implementation

Imaging Backbone: F_ψ is a 2D UNet [22] (without the output classification layer) to extract features $\mathbf{Z}$. The four modalities are concatenated as the input (with 4 channels) to F_ψ. For a fair comparison, we use this same UNet implementation as the backbone for all models.

vMFNet and vMFBrain[1]: T_θ is a shallow convolutional network. W_ω is a classifier model. Following [11], we set the variance of the vMF distributions as 30. The number of kernels is set to 8, as this number performed best empirically in our early experiments. For vMF kernel initialisation, we pre-train a 2D UNet for 10 epochs to reconstruct the input image with all the training data. After training, we extract the corresponding feature vectors and perform k-means clustering, then use the discovered cluster centres to initialise the vMF kernels.

Training: All models are implemented in PyTorch [21] and are trained using an NVIDIA 3090 GPU. Models are trained using the Adam optimiser [10] with a learning rate of $1 \times e^{-4}$ using batch size 32. In semi-supervised and weakly supervised settings, we consider the use of different percentages of fully labelled data to train the models. For this purpose, we randomly sample 2D image slices and the corresponding pixel-level labels from the whole training dataset.

Table 1. Dice (%) and Hausdorff Distance (HD) results for the task of **whole tumour segmentation**. We report the mean and standard deviation across volumes.

Metrics	Dice (↑)				HD (↓)			
Pixel labels	0.1%	0.5%	1%	100%	0.1%	0.5%	1%	100%
UNet	80.66_{10}	$86.39_{7.7}$	$87.34_{7.0}$	$90.84_{5.6}$	9.18_{10}	$6.60_{8.1}$	7.37_{10}	$\mathbf{4.49_{7.2}}$
SDNet	79.20_{11}	$86.38_{7.6}$	$87.96_{6.6}$	$\mathbf{90.96_{5.3}}$	11.85_{13}	$7.24_{9.3}$	$6.11_{8.4}$	$4.87_{8.3}$
vMFNet	$81.30_{9.6}$	$86.14_{7.8}$	$87.98_{6.6}$	$90.62_{5.8}$	11.62_{13}	9.12_{12}	$7.15_{9.6}$	$5.20_{8.2}$
vMFBrain w/o weak	79.70_{10}	$84.92_{8.1}$	$87.26_{6.7}$	$90.67_{5.8}$	13.89_{14}	9.80_{13}	$7.18_{9.4}$	$4.93_{7.3}$
vMFBrain	$\mathbf{85.64_{7.8}}$	$\mathbf{88.64_{6.8}}$	$\mathbf{89.04_{6.7}}$	$90.58_{5.6}$	$\mathbf{7.75_{7.8}}$	$\mathbf{6.18_{8.6}}$	$\mathbf{6.14_{8.4}}$	$4.60_{6.5}$

Table 2. Dice (%) and Hausdorff Distance (HD) results for the task of **tumour sub-region segmentation**. We report the mean and standard deviation across volumes.

0.1% pixel labelled data	ED		ET		NE	
	Dice (↑)	HD (↓)	Dice (↑)	HD (↓)	Dice (↑)	HD (↓)
UNet	71.47_{12}	9.60_{11}	$83.74_{8.4}$	$\mathbf{5.19_{5.7}}$	79.42_{10}	$10.24_{7.9}$
SDNet	75.87_{11}	10.17_{11}	$82.45_{8.8}$	7.74_{12}	$80.70_{9.7}$	9.89_{11}
vMFNet	71.11_{12}	$10.06_{9.8}$	$80.92_{9.6}$	7.99_{11}	78.37_{11}	12.97_{11}
vMFBrain w/o weak	70.65_{13}	15.65_{15}	79.36_{12}	13.13_{17}	$79.33_{9.8}$	$9.50_{9.3}$
vMFBrain w/ whole tumour weak	75.02_{11}	11.56_{12}	$84.59_{8.5}$	8.17_{12}	$79.48_{9.6}$	$10.02_{9.1}$
vMFBrain w/ tumour sub-region weak	$\mathbf{78.43_{9.8}}$	$\mathbf{9.14_{8.8}}$	$\mathbf{85.77_{8.2}}$	$5.90_{7.7}$	$\mathbf{81.31_{9.0}}$	$\mathbf{8.08_{7.0}}$

---

[1] The code for vMFNet is available at https://github.com/vios-s/vMFNet.

4.4 Results

We compare model performance quantitatively using volume-wise Dice (%) and Hausdorff Distance (95%) (HD) [8] as the evaluation metrics, and qualitatively using the interpretability and compositionality of representations. In Table 1 and Table 2, for semi-supervised and weakly supervised approaches, the training data contains all unlabelled or weakly labelled data alongside different percentages of fully labelled data. UNet is trained with different percentages of labelled data only. Bold numbers indicate the best performance. Arrows (↑, ↓) indicate the direction of metric improvement.

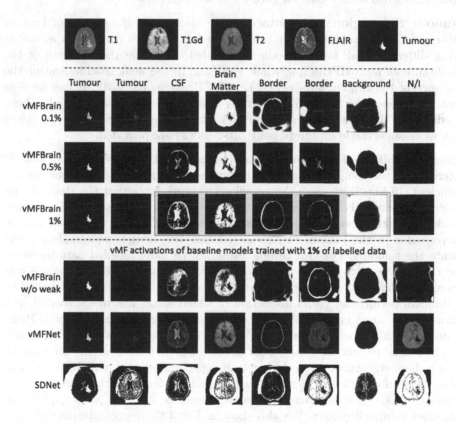

Fig. 2. Visualisation of vMF compositional representations (whole tumour supervision). We show the 4 input image modalities, the ground truth tumour segmentation mask, and all 8 vMF channels for the vMFBrain and baseline models trained with different percentages of labelled data. In the red boxes, the other interpretable vMF activations (excluding the tumour kernels) are highlighted. The vMF channels are ordered manually. For the vMFBrain channels, we label with a clinician's visual interpretation of which image features activated each kernel. N/I denotes non-interpretable. (Color figure online)

Brain Tumour Segmentation with Weak Labels: Overall, as reported in Table 1, the proposed vMFBrain model achieves best performance for most of the cases, particularly when very few annotations are available, i.e. the 0.1% case. When dropping the weak supervision (vMFBrain w/o weak), we observe reduced performance, which confirms the effectiveness of weak supervision. We also observe that the reconstruction of the original image (in vMFNet) does not help. It is possible that reconstruction of the tumour does not help here because the tumour has inconsistent appearance and location between different scans. With more annotated data, all models gradually achieve better performance, as expected. Notably, with only 1% labelled data vMFBrain achieves comparable performance (89.04 on Dice and 6.14 on HD) to the fully supervised UNet trained with all labelled data (90.84 on Dice and 4.49 on HD).

Tumour Sub-region Segmentation: We also report the results of tumour sub-region segmentation task in Table 2. For this task, we perform experiments using different weak labels: a) the weak label indicating the presence of the whole tumour i.e. vMFBrain w/whole weak and b) the weak label indicating the presence of the tumour sub-regions i.e. vMFBrain w/sub weak. It can be seen that our proposed vMFBrain performed best with both types of weak labels. Predictably, the best performance occurs when more task-specific weak labels (i.e. weak supervision on the tumour sub-regions) are provided.

Interpretability of Compositional Representations: We are particularly interested in the compositionality of the representations when pixel labels are not sufficient. In Fig. 2, we show the kernel activations. Note that the channels are ordered manually. For different runs, the learning is emergent such that kernels randomly learn to represent different components. Clearly, one of the kernels corresponds to the tumour in all cases. Using this kernel, we can detect and locate the tumours. For vMFBrain, training with more labelled data improves the compositionality of the kernels and the activations i.e. different kernels correspond to different anatomical or pathological structures, which are labelled by a clinician performing visual inspection of which image features activated each channel. The most interpretable and compositional representation is vMFBrain trained with 1% labelled data. As highlighted in the red boxes, the kernels relate to CSF, brain matter, and the border of the brain even without any information about these structures given during training. Qualitatively, vMFBrain decomposes this information better into each kernel i.e. learns better compositional representations compared to other baseline models. Notably, weak supervision improves compositionality. We also show in Fig. 3 the representations for sub-region segmentation. Overall, we observe that with the more task-specific weak labels, the kernels learn to be more aligned with the sub-region segmentation task, where less information on other clinically relevant features is learnt.

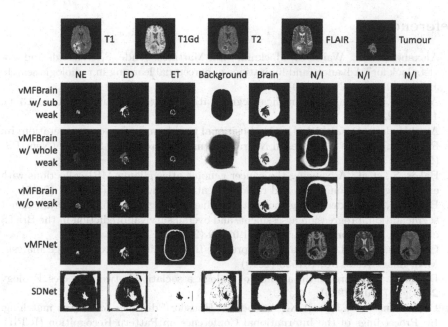

Fig. 3. Visualisation of vMF compositional representations (tumour sub-region supervision). We show the 4 input image modalities, the corresponding ground truth tumour sub-region segmentation mask and all 8 channels of the representations for the models trained with 1% of labelled data. The channels are ordered manually. For the vMF-Brain channels, we label with a clinician's visual interpretation of which image features activated each kernel. N/I denotes non-interpretable.

5 Conclusion

In this paper, we have presented vMFBrain, a compositional representation learning framework. In particular, we constructed weak labels indicating the presence or absence of the brain tumour and tumour sub-regions in the image. Training with weak labels, better compositional representations can be learnt that produce better brain tumour segmentation performance when the availability of pixel-level annotations is limited. Additionally, our experiments show the interpretability of the compositional representations, where each kernel corresponds to specific anatomical or pathological structures. Importantly, according to our experiments and the results reported in previous studies [15,18], the vMF-based compositional representation learning framework is robust and applicable to different medical datasets and tasks. In future work, we might consider transferring vMFBrain to 3D in order to process wider spatial context for each structure.

Acknowledgements. S.A. Tsaftaris acknowledges the support of Canon Medical and the Royal Academy of Engineering and the Research Chairs and Senior Research Fellowships scheme (grant RCSRF1819n8n25). Many thanks to Patrick Schrempf and Joseph Boyle for their helpful review comments.

References

1. Alexander, R.G., Waite, S., Macknik, S.L., Martinez-Conde, S.: What do radiologists look for? Advances and limitations of perceptual learning in radiologic search. J. Vis. **20**(10), 17–17 (2020)
2. Antonelli, M., et al.: The medical segmentation decathlon. Nat. Commun. **13**(1), 4128 (2022)
3. Arad Hudson, D., Zitnick, L.: Compositional transformers for scene generation. In: Proceedings of the Advances in Neural Information Processing Systems (NeurIPS) (2021)
4. Bakas, S., et al.: Advancing the cancer genome atlas glioma MRI collections with expert segmentation labels and radiomic features. Sci. Data **4**(1), 1–13 (2017)
5. Bakas, S., et al.: Identifying the best machine learning algorithms for brain tumor segmentation, progression assessment, and overall survival prediction in the BraTS challenge. arXiv preprint arXiv:1811.02629 (2018)
6. Chartsias, A., et al.: Disentangled representation learning in cardiac image analysis. Med. Image Anal. **58**, 101535 (2019)
7. Dice, L.R.: Measures of the amount of ecologic association between species. Ecology **26**(3), 297–302 (1945)
8. Dubuisson, M.P., Jain, A.K.: A modified Hausdorff distance for object matching. In: Proceedings of the International Conference on Pattern Recognition (ICPR), vol. 1, pp. 566–568. IEEE (1994)
9. Huynh, D., Elhamifar, E.: Compositional zero-shot learning via fine-grained dense feature composition. In: Proceedings of the Advances in Neural Information Processing Systems (NeurIPS), vol. 33, pp. 19849–19860 (2020)
10. Kingma, D.P., Ba, J.: Adam: a method for stochastic optimization. In: Proceedings of the International Conference on Learning Representations (ICLR) (2015)
11. Kortylewski, A., He, J., Liu, Q., Yuille, A.L.: Compositional convolutional neural networks: a deep architecture with innate robustness to partial occlusion. In: Proceedings of the IEEE/CVF Conference on Computer Vision and Pattern Recognition (CVPR), pp. 8940–8949 (2020)
12. Kortylewski, A., Liu, Q., Wang, H., Zhang, Z., Yuille, A.: Combining compositional models and deep networks for robust object classification under occlusion. In: Proceedings of the IEEE/CVF Winter Conference on Applications of Computer Vision (CVPR), pp. 1333–1341 (2020)
13. Lake, B.M., Salakhutdinov, R., Tenenbaum, J.B.: Human-level concept learning through probabilistic program induction. Science **350**(6266), 1332–1338 (2015)
14. Liu, N., Li, S., Du, Y., Tenenbaum, J., Torralba, A.: Learning to compose visual relations. In: Proceedings of the Advances in Neural Information Processing Systems (NeurIPS), vol. 34 (2021)
15. Liu, X., Sanchez, P., Thermos, S., O'Neil, A.Q., Tsaftaris, S.A.: Compositionally equivariant representation learning. arXiv preprint arXiv:2306.07783 (2023)
16. Liu, X., Thermos, S., Chartsias, A., O'Neil, A., Tsaftaris, S.A.: Disentangled representations for domain-generalized cardiac segmentation. In: Proc. International Workshop on Statistical Atlases and Computational Models of the Heart (STACOM). pp. 187–195 (2020)
17. Liu, X., Thermos, S., O'Neil, A., Tsaftaris, S.A.: Semi-supervised meta-learning with disentanglement for domain-generalised medical image segmentation. In: de Bruijne, M., et al. (eds.) MICCAI 2021. LNCS, vol. 12902, pp. 307–317. Springer, Cham (2021). https://doi.org/10.1007/978-3-030-87196-3_29

18. Liu, X., Thermos, S., Sanchez, P., O'Neil, A.Q., Tsaftaris, S.A.: vMFNet: compositionality meets domain-generalised segmentation. In: Wang, L., Dou, Q., Fletcher, P.T., Speidel, S., Li, S. (eds.) MICCAI 2022. LNCS, vol. 13437, pp. 704–714. Springer, Cham (2022)
19. Menze, B.H., et al.: The multimodal brain tumor image segmentation benchmark (BraTS). IEEE Trans. Med. Imaging **34**(10), 1993–2024 (2014)
20. Milletari, F., Navab, N., Ahmadi, S.A.: VNet: fully convolutional neural networks for volumetric medical image segmentation. In: 3DV, pp. 565–571. IEEE (2016)
21. Paszke, A., et al.: Pytorch: an imperative style, high-performance deep learning library. In: Proceedings of the Advances in Neural Information Processing Systems (NeurIPS), vol. 32 (2019)
22. Ronneberger, O., Fischer, P., Brox, T.: U-Net: convolutional networks for biomedical image segmentation. In: Navab, N., Hornegger, J., Wells, W.M., Frangi, A.F. (eds.) MICCAI 2015. LNCS, vol. 9351, pp. 234–241. Springer, Cham (2015). https://doi.org/10.1007/978-3-319-24574-4_28
23. Singh, K.K., Ojha, U., Lee, Y.J.: Finegan: unsupervised hierarchical disentanglement for fine-grained object generation and discovery. In: Proceedings of the IEEE/CVF Conference on Computer Vision and Pattern Recognition (CVPR), pp. 6490–6499 (2019)
24. Thermos, S., Liu, X., O'Neil, A., Tsaftaris, S.A.: Controllable cardiac synthesis via disentangled anatomy arithmetic. In: de Bruijne, M., et al. (eds.) MICCAI 2021. LNCS, vol. 12903, pp. 160–170. Springer, Cham (2021). https://doi.org/10.1007/978-3-030-87199-4_15
25. Tokmakov, P., Wang, Y.X., Hebert, M.: Learning compositional representations for few-shot recognition. In: Proceedings of the IEEE/CVF Conference on Computer Vision and Pattern Recognition (CVPR), pp. 6372–6381 (2019)
26. Yuan, X., Kortylewski, A., et al.: Robust instance segmentation through reasoning about multi-object occlusion. In: Proceedings of the IEEE/CVF Conference on Computer Vision and Pattern Recognition (CVPR), pp. 11141–11150 (2021)
27. Yuille, A.L., Liu, C.: Deep nets: what have they ever done for vision? Int. J. Comput. Vision **129**, 781–802 (2021)
28. Zhang, Y., Kortylewski, A., Liu, Q., et al.: A light-weight interpretable compositionalnetwork for nuclei detection and weakly-supervised segmentation. arXiv preprint arXiv:2110.13846 (2021)

Hierarchical Compositionality in Hyperbolic Space for Robust Medical Image Segmentation

Ainkaran Santhirasekaram[1]([✉]), Mathias Winkler[2], Andrea Rockall[2], and Ben Glocker[1]

[1] Department of Computing, Imperial College London, London, UK
a.santhirasekaram19@ic.ac.uk
[2] Department of Surgery and Cancer, Imperial College London, London, UK

Abstract. Deep learning based medical image segmentation models need to be robust to domain shifts and image distortion for the safe translation of these models into clinical practice. The most popular methods for improving robustness are centred around data augmentation and adversarial training. Many image segmentation tasks exhibit regular structures with only limited variability. We aim to exploit this notion by learning a set of base components in the latent space whose composition can account for the entire structural variability of a specific segmentation task. We enforce a hierarchical prior in the composition of the base components and consider the natural geometry in which to build our hierarchy. Specifically, we embed the base components on a hyperbolic manifold which we claim leads to a more natural composition. We demonstrate that our method improves model robustness under various perturbations and in the task of single domain generalisation.

1 Introduction

The task of developing robust medical image segmentation models is challenging due to the variation in acquisition protocols across multiple source domains leading to different textural profiles [8]. Additionally, the complex acquisition process also leads to artefacts such as noise and motion.

There have been various approaches in the literature to address distributional shifts from the training to the test phase. The most common approach is to augment the training data with synthetic perturbations of the input space [10, 26, 27, 30]. Adversarial training [6, 18, 24] schemes and self-supervised learning strategies [4] have also been proposed to improve model robustness.

The spatial variability of segmentation maps across medical images is limited due to structural consistency. Therefore, one can exploit this notion by limiting

---

Supplementary Information The online version contains supplementary material available at https://doi.org/10.1007/978-3-031-45857-6_6.

L. Koch et al. (Eds.): DART 2023, LNCS 14293, pp. 52–62, 2024.
https://doi.org/10.1007/978-3-031-45857-6_6

the search space when learning the embeddings of a deep learning based segmentation model. We aim to achieve this by learning a fixed set of base components in the latent space whose combinatorial composition can represent the entire training and test distribution. The base components are learnt by discretising the latent space via quantisation to form a dictionary of components for each class [25]. In order to learn the composition of the sampled based components in a hierarchical manner, we embed our base components in hyperbolic space. It has recently been shown that spatial representations in the brain exhibit hyperbolic geometry [29]. We hypothesise hyperbolic geometry will provide the natural structure to efficiently embed the base components without distortion.

The contributions of this paper are summarized as follows:

- We improve the robustness of deep learning based segmentation models by imposing hierarchical compositionality in the latent space.
- This is the first work to consider the curvature upon which to learn a dictionary of base components via quantisation. Here, we consider negative curvature (Poincare space) to best induce hierarchical compositionality.

2 Background: Hyperbolic Geometry

We will first introduce some important notions of a manifold. One can define a d dimensional *manifold* $\mathcal{M}$ as a topological space whereby a local neighbourhood around each point is homeomorphic to an open subset of Euclidean space. The *tangent Space* $\mathcal{T}_x\mathcal{M}$ of a point x on a differentiable manifold is a vector space whose elements are tangent vectors of all possible paths on manifold passing through x. A manifold equipped with a Riemmanian metric tensor which allows to measure geometric properties such as distances and angles is known as a *Riemmanian manifold*. The *Riemmanian metric tensor* defines the set of inner products $g_x : \mathcal{T}_x\mathcal{M} \times \mathcal{T}_x\mathcal{M} \to \mathbb{R}$ of every point x on $\mathcal{M}$.

The curvature of a manifold is captured in the metric tensor of the manifold. Hence, curvature governs the geometric properties of the manifold. We consider manifolds with negative curvature which are defined by the properties of hyperbolic geometry. There are multiple models of hyperbolic geometry but for the purpose of this work, we will focus on the Poincare unit ball. Formally, the d dimensional Poincare ball, $\mathbb{B}$ with unit radius and -1 curvature is formally defined as: $\mathbb{B}^{d,1} = \{x \in \mathbb{R}^d | \|x\| < 1\}$. In order to use the Poincare model for representation learning, it is important to define how to measure the geodesic distances in Poincare space. The Riemannian metric tensor for the Poincare unit ball is calculated as $g_x^{\mathbb{B},1}(u,v) = \left(\frac{2}{1-\|x\|^2}\right)^2 (u.v)$. We can then derive an equation for the geodesic distance between two points, $(u, v) \in \mathbb{B}^{d,1}$ below.

$$d^{\mathbb{B},1}(u,v) = \cosh^{-1}\left(1 + 2\left(\frac{\|u-v\|^2}{(1-\|u\|^2)(1-\|v\|^2)}\right)\right) \qquad (1)$$

Eq. 1 demonstrates straight lines in Poincare space as arcs of circles that are orthogonal to the circle boundary. Additionally, the metric tensor tells us that the volume in Poincare space grows exponentially from the origin [5]. This is a useful property for learning tree-structured hierarchical representations with a high branching factor compared to in Euclidean space which grows quadratically.

Next, we define a unique mapping between the Poincare unit ball and Euclidean space defined in the tangent space of the manifold. In this work, we map to the tangential plane at the origin, o. The first equation in 2 defines an exponential mapping, $exp_o^{\mathbb{B},1}(y)$ which is a homeomorphic map from the point y in Euclidean space, $y \in \mathcal{T}_o\mathbb{B}^{d,1}$ to the Poincare unit ball, $v \in \mathbb{B}^{d,1}$. The logarithmic map, $log_o^{\mathbb{B},1}(v)$ shown in the second equation in 2 is the inverse map of the exponential map.

$$exp_o^{\mathbb{B},1}(y) = \left(tanh(\|y\|_2)\frac{y}{\|y\|_2} \right), \quad log_o^{\mathbb{B},1}(v) = \left(0, arctanh(\|v\|_2)\frac{v}{\|v\|_2} \right) \quad (2)$$

3   Related Work

The goal of single domain generalisation (SDG) is to generalise well to target domains after training on one source domain. Strong data augmentation strategies such as BigAug [30], CutOut [10] and MixUp [28] provide the most direct solution to improve domain generalisability of segmentation models. The more recent approaches have used adversarial methods such as AdvBias [6] which learns to generate bias field deformations of MRI images to improve model robustness. An interesting approach to tackle SDG segmentation is to learn textural invariant features. RandConv [27] attempts to achieve this by incorporating a randomised convolution layer in the input space. JiGen [4] is a self-supervised method which provides another natural choice for SDG.

The idea of compositionality has been incorporated into neural networks to improve the robustness of image classification under partial occlusion [14]. Recently, compositional representational learning by learning a discrete latent space has shown to improve the robustness of segmentation models [17,23].

In our work, hyperbolic embeddings provide the natural basis for hierarchically composing a discretised latent space. Hyperbolic embeddings have become increasingly popular for dealing with hierarchical data such as graphs [5,16] and can even informally be expressed as the continuous version of trees. For example, [20] suggests that the Poincare model is better equipped to learn word embeddings to simultaneously capture a hierarchy and similarity between embeddings, which led to promising results in link prediction. Recently, hyperbolic image segmentation has been developed to classify output pixels in hyperbolic space by learning hyperbolic planes for improved class separation [1].

4   Methods

4.1   Base Component Learning

In our work, we assume that a segmentation network can be decomposed into an encoder (Φ_e) to map the input space, X to a lower dimensional embedding space denoted, $\mathcal{E}$ and a decoder to map $\mathcal{E}$ to the output space, Y.

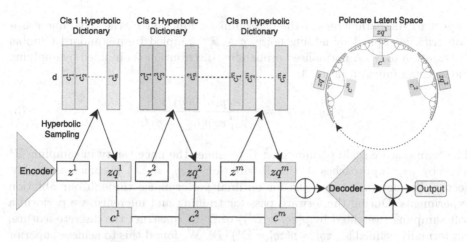

Fig. 1. Our method (HypC) learns base dictionaries embedded in hyperbolic space for each class. The embedding features, z are split into m (number of classes) groups with each group sampling its respective class hyperbolic dictionary, $\mathbb{D}^i$ to form zq^i which is composed to form class segmentation outputs, c^i.

We propose to discretise $\mathcal{E}$ into a set of N base components to form a d dimensional dictionary, $\mathbb{D}$ using quantisation [25]. However, in order to induce compositionality into our base components, we enforce $\mathbb{D}$ to be divided into m classes such that $\mathbb{D}^i$ contains the base components required to compose each class segmentation output as shown in Fig. 1. This is further enforced by using only grouped convolutions equivalent to the number of classes in the encoder and decoder. The embedding features, z are divided into m groups with each group of n features, z^i sampled from its respective class dictionary, $\mathbb{D}^i$ to produce discrete features, zq^i. zq^i represents the components required to compose lower dimensional class segmentation outputs, c^i with d dimensionality. This allows us to directly enforce a hierarchical relationship between zq^i and c^i in d dimensional hyperbolic space. The composition of the features in zq^i to form c^i shown in Fig. 1 is performed with a 1×1 convolutional layer.

We propose to discretise the embedding space with Gumbel quantisation [12]. A random variable, G is defined to have the standard Gumbel distribution if it satisfies, $G = -log(-log(U))$ given U is sampled uniformly in the closed interval, $[0, 1]$. The Gumbel softmax trick can be used to sample a categorical distribution which in our case is the dictionary $\mathbb{D}^i$ in a fully differentiable manner [12]. Equation 3, shows a categorical distribution for each discrete feature, $zq_l^i \in zq^i$ over the components in $\mathbb{D}^i$. This essentially tells us the probability of sampling each of the n components in $\mathbb{D}^i$ to form zq_l^i, denoted $p(zq_l^i = \mathbb{D}^i) \in \mathbb{R}^n$. In Eq. 3, the probability distribution is parameterised by $\alpha_l^i \in \mathbb{R}^n$ which one can note are unnormalised probabilities proportional to $log(P(zq_l^i = \mathbb{D}^i))$. α_l^i is determined via a parameterised linear transformation, F of z_l^i to match the number dictionary elements (n) followed by a ReLU activation function. The parameters are shared across features in z^i. However, in order to compute the gradient with respect to the parameters of the deterministic function F, the reparameterisation trick is applied by adding noise, $g \in \mathbb{R}^n$ sampled from standard Gumbel distribution to α_l^i. A normalised probability distribution is obtained by applying the softmax function in Eq. 3.

$$p(zq_l^i = \mathbb{D}^i) = \frac{exp(\alpha_l^i + g)/\tau}{\sum_{j=1}^{j=n} exp(\alpha_{l,j}^i + g)/\tau} \tag{3}$$

The temperature scale parameter, τ determines the uncertainty in sampling $\mathbb{D}^i$ whereby as τ approaches 0, the distribution $p(zq_l^i = \mathbb{D}^i)$ becomes a one hot vector. We determine $\tau = 0.5$ for optimal performance through our ablation experiments. During the forward pass for training and inference, we perform a soft sampling (weighted by $p(zq_l^i = \mathbb{D}^i)$) of $\mathbb{D}^i$ to construct the discrete feature, zq_l^i formally defined as, $zq_l^i = p(zq_l^i = \mathbb{D}^i)^\top \mathbb{D}^i$. We found this to achieve superior results compared to performing a hard sampling of $\mathbb{D}^i$ via the $argmax_j$ function.

4.2   Hyperbolic Embeddings

We initialise $\mathbb{D}^i$ on the Poincare unit ball ($\mathbb{B}^{d,1}$) but perform soft Gumbel quantisation on the tangent plane of the origin via a logarithmic mapping of $\mathbb{D}^i$ formally defined as: $zq_l^i = p(zq_l^i = \mathbb{D}^i)^\top log_o^{\mathbb{B},1}(\mathbb{D}^i)$.

Equation 2 is then used to apply an exponential mapping ($exp_o^{\mathbb{B},1}$) to $\mathbb{D}^i$ to move back into Poincare space.

An exponential mapping (Eq. 2) of zq and c to the Poincare unit ball will exploit the natural geometry of hyperbolic manifolds to enforce a hierarchical composition of zq to form c as shown in the Poincare latent space in Fig. 1. This is achieved by defining a loss function shown in Eq. 4. Equation 4 shows the p features in zq^i are being pulled closer together in Poincare distance (Eq. 1) to its composition, c^i while being pushed apart from the compositions not composed from zq^i defined as $\bar{c}^i$. zq is then mapped back to euclidean space with a logarithmic mapping and inputted into the decoder to construct the high dimensional segmentation output (output 1) shown in Fig. 1. We predict this hyperbolic hierarchical organisation of the latent space serves to learn more robust embedding features for constructing the segmentation output.

$$\mathcal{L}_{Poincare} = \frac{\sum_{i=1}^{i=m} \sum_{l=1}^{l=p} e^{d^{\mathbb{B},1}(zq_l^i, c^i)}}{\sum_{i=1}^{i=m} \sum_{l=1}^{l=p} e^{d^{\mathbb{B},1}(zq_l^i, \bar{c}^i)} + \sum_{i=1}^{i=m} \sum_{i'=1}^{i'=m-1} e^{d^{\mathbb{B},1}(c^i, \bar{c}^i)}} \tag{4}$$

We apply a dice loss between output 1; y' and the label, y as well as between output 2; y_2' and a down-sampled label of y to match the dimensionality of y_2' denoted, y_2 (Fig. 1). The total loss during training is calculated as:
$$\mathcal{L}_{Total} = \mathcal{L}_{dice}(y', y) + \mathcal{L}_{dice}(y_2', y_2) + \mathcal{L}_{Poincare}.$$

5   Experiments

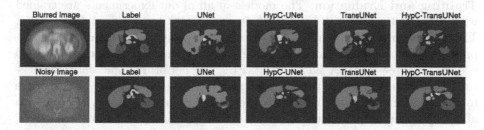

Fig. 2. The figure shows the segmentations of an abdominal CT slice with 20% Gaussian noise addition (bottom row) and Gaussian blur (top row) produced by the UNet and TransUNet before and after incorporating our method.

Table 1. The mean dice score and Haussdorf distance ± standard deviation before and after hierarchical hyperbolic compositionality(HypC) is applied to 2 segmentation models under various perturbations in the input space.

	Baseline	Gauss	S&P	Blur	Motion	Contrast	Intensity
	Dice						
UNet	.77±.08	.71±.10	.69±.06	.72±.11	.63±.13	.62±.14	.68±.11
HypC-UNet	**.78±.07**	**.75±.04**	**.72±.04**	**.76±.10**	**.67±.04**	**.64±.07**	**.71±.11**
TransUNet	.82±.08	.78±.12	.76±.09	.79±.08	.70±.11	.74±.17	.73±.11
HypC-TransUNet	**.83±.11**	**.81±.09**	**.79±.10**	**.81±.03**	**.74±.15**	**.76±.11**	**.75±.13**
	Haussdorf Distance						
UNet	.24±.12	.38±.15	.40±.19	.30±.16	.45±.12	.44±.20	.37±.18
HypC-UNet	**.22±.10**	**.34±.13**	**.38±.19**	**.25±.10**	**.39±.18**	**.39±.16**	**.34±.13**
TransUNet	.23±.08	.33±.15	.35±.16	.28±.10	.34±.08	.33±.15	.30±.09
HypC-TransUNet	**.22±.04**	**.30±.08**	**.33±.16**	**.25±.13**	**.32±.12**	**.27±.10**	**.27±.15**

5.1   Datasets and Training

Abdomen: This abdominal dataset is made up of 30 CT scans with 13 labels from a single domain as part of the Beyond the Cranial Vault (BTCV) dataset [15]. During training and inference, we randomly crop $96 \times 96 \times 96$ patches which are normalised between 0 and 1.

Prostate: The prostate MRI dataset consists of 60 T2 weighted scans with 2 labels from the NCI-ISBI13 Challenge [2,3,9]. This dataset is equally split into 2 different domains. The first domain is from the Boston Medical Centre (BMC) where images are acquired on a 1.5T scanner with an endorectal coil.

The second domain acquires images on a 3T scanner with a surface coil [3] in the Radboud University Nijmegen Medical Centre (RUNMC). All images are normalised between 0 and 1 and centre cropped to $256 \times 256 \times 24$.

Training and Evaluation: The models in all of our experiments are trained with Adam optimisation [13] at a learning rate of 0.0001 and weight decay (0.05). The parameters in $\mathbb{D}^i$ are mapped to Euclidean space using a logarithmic mapping (Eq. 2) for optimisation. All models were trained for a maximum of 500 epochs on three NVIDIA RTX 2080 GPUs. We perform simple data augmentation with random flipping (horizontal and vertical) and rotation. We evaluate model performance with the Dice score and Hausdorff distance. We perform ablations to learn dictionaries, $\mathbb{D}^i$ which are as sparse as possible without affecting segmentation performance in order to improve model robustness. Please refer to supplementary material for the ablation study results and other experiments.

Fig. 3. Segmentation of a prostate slice from the BMC dataset when training on the RUNMC dataset with different SDG methods and our method.

5.2   Perturbation Experiments

In the first experiment, we observe the effect of incorporating a hyperbolic hierarchical compositional structure (HypC) in the embedding space of a segmentation model under various types of perturbations. We specifically analyse the robustness of the 3D UNet [19,22] and TransUNet [7], two popular segmentation architectures for the Abdomen dataset split into 24 training and 6 test samples. In both models, 3D pre-activation residual blocks [11] make up each level of a 5-level, convolutional encoder and decoder. In the TransUNet [7], the transformer block consisting of 12 layers and 8 heads is incorporated after the HypC block.

Noise levels were varied between 1 to 30 % for the Gaussian and Salt and Pepper (S&P) noise experiments. We use a Gaussian kernel with sigma values ranging from 0.1 to 2.0 to distort images with Gaussian blur. The contrast change pertubations involved adjusting the gamma value between 0.5 and 4.5 while random motion is applied using the TorchIO deep learning library [21]. Intensity values are scaled by a factor between 0.8 and 1.2.

Table 1 highlights the improved performance of both the UNet and TransUNet under all types of perturbation effects after imposing HypC into the embedding space. We specifically noticed improve robustness of the two models under both textural and spatial distortions such as noise and motion perturbation. We highlight our findings visually in Fig. 2. Here, one can appreciate the smoother boundaries and a reduced number of unconnected components after incorporating our method into the UNet under Gaussian noise and blur addition. Please refer to supplementary material for further results on the importance of curvature to impose hierarchical compositionality for robust segmentation.

Table 2. The average dice score and Hausdorff distance (HD) ± standard deviations using several single domain generalisation methods compared to our approach under different geometric structures (curvatures)

	$BMC \rightarrow RUNMC$		$RUNMC \rightarrow BMC$	
	Dice	HD	Dice	HD
Baseline	.45±.11	.44±.17	.52±.18	.40±.11
CutOut [10]	.50±.12	.38±.15	.53±.15	.35±.09
AdvBias [6]	.54±.16	.30±.16	.56±.11	.31±.10
RandConv [27]	.56±.18	.35±.14	.59±.22	.30±.11
BigAug [30]	**.58±.14**	.29±.09	**.63±.20**	.25±.10
Jigen [4]	.51±.19	.42±.16	.53±.10	.31±.05
VQ-UNet [23]	.53±.14	.34±.20	.57±.15	.32±.15
Ours (Euclidean)	.55 ±.17	.31±.12	.58±.16	.27±.11
Ours (Hyper-spherical)	.57±.14	.32±.09	.60±.13	.28±.14
Ours (Hyperbolic)	**.58±.08**	**.28±.12**	.61±.16	**.24±.11**

5.3   Single Domain Generalisation

We use the prostate dataset in our SDG experiments where we train on the RUNMC dataset to test on the BMC dataset and vice versa. In order to accommodate the anisotropic nature of the prostate MRI images, we use a hybrid 2D/3D UNet as our baseline model. This models has 4 levels, each level consisting of 2, 2D pre-activation residual blocks [11] followed by a 2D strided convolutional layer for the encoder and a 2D up-sampling block for the decoder. The bottleneck level consists of a 3D pre-activation residual block. The hyperbolic hierarchical compositional structure is built into the bottleneck of the baseline model. The importance of curvature in our method is assessed by integrating Euclidean and hyper-spherical(+1 curvature) hierarchical compositional blocks into our baseline model. We compare to the SDG methods; RandConv [27], CutOut [10], BigAug [30], AdvBias [6] and Jigen [4] applied to the baseline. We also compare to the VQ-UNet [23] which learns a discrete latent space via vector quantisation.

In Table 2, we demonstrate the superior performance of our approach in both metric scores compared to most other SDG methods. We achieve similar performance to BigAug [30] highlighting how incorporating a compositional structure under the natural geometric constraints improves the generalisability of a segmentation model even without an aggressive augmentation pipeline. This is demonstrated with a visual example in Fig. 3 which shows the more anatomically plausible segmentations maps produced by our method. Additionally, improved segmentation performance is noted in Table 2 by naturally embedding the base components on a hyperbolic manifold as opposed to on a hyper-sphere or in Euclidean space in order to instill hierarchical compositionality.

6   Conclusion

We propose to integrate hierarchical compositionality equipped with the natural geometry of a hyperbolic manifold into any existing segmentation model for improved robustness. Specifically, our approach improves model robustness to textural and spatial perturbations as well as domain shifts. We highlight in our results, it is better to hierarchically compose and sample a set of learnt base components embedded in hyperbolic space compared to in spherical or Euclidean space. The main limitation of this work is that we assume a fixed unit negative curvature for all class codebooks to impose our compositionality constraint. Therefore, in future work, we aim to explore learnable curvature for compositionality in the embedding space to improve performance of our method.

Acknowledgements. This work was supported and funded by Cancer Research UK (CRUK) (C309/A28804).

References

1. Atigh, M.G., Schoep, J., Acar, E., van Noord, N., Mettes, P.: Hyperbolic image segmentation. In: Proceedings of the IEEE/CVF Conference on Computer Vision and Pattern Recognition, pp. 4453–4462 (2022)
2. Bloch, N., et al.: Cancer imaging archive wiki. https://doi.org/10.7937/K9/TCIA. 2015.zF0vlOPv (2015)
3. Bloch, N., Madabhushi, A., Huisman, H., Freymann, J., Kirby, J., Grauer, M., Enquobahrie, A., Jaffe, C., Clarke, L., Farahani, K.: Nci-isbi 2013 challenge: automated segmentation of prostate structures. Cancer Imaging Archive **370**, 6 (2015)
4. Carlucci, F.M., D'Innocente, A., Bucci, S., Caputo, B., Tommasi, T.: Domain generalization by solving jigsaw puzzles. In: Proceedings of the IEEE/CVF Conference on Computer Vision and Pattern Recognition, pp. 2229–2238 (2019)
5. Chami, I., Ying, Z., Ré, C., Leskovec, J.: Hyperbolic graph convolutional neural networks. Advances in neural information processing systems 32 (2019)
6. Chen, C., Qin, C., Qiu, H., Ouyang, C., Wang, S., Chen, L., Tarroni, G., Bai, W., Rueckert, D.: Realistic adversarial data augmentation for mr image segmentation. In: International Conference on Medical Image Computing and Computer-Assisted Intervention, pp. 667–677. Springer (2020)
7. Chen, J., et al.: Transunet: Transformers make strong encoders for medical image segmentation. arXiv preprint arXiv:2102.04306 (2021)
8. Chen, Y.: Towards to robust and generalized medical image segmentation framework. arXiv preprint arXiv:2108.03823 (2021)
9. Clark, K., et al.: The cancer imaging archive (TCIA): maintaining and operating a public information repository. J. Digit. Imaging **26**, 1045–1057 (2013)

10. DeVries, T., Taylor, G.W.: Improved regularization of convolutional neural networks with cutout. arxiv 2017. arXiv preprint arXiv:1708.04552 (2017)
11. He, K., Zhang, X., Ren, S., Sun, J.: Identity mappings in deep residual networks. In: Computer Vision-ECCV 2016: 14th European Conference, Amsterdam, The Netherlands, October 11–14, 2016, Proceedings, Part IV 14, pp. 630–645. Springer (2016)
12. Jang, E., Gu, S., Poole, B.: Categorical reparameterization with gumbel-softmax. arXiv preprint arXiv:1611.01144 (2016)
13. Kingma, D.P., Ba, J.: Adam: A method for stochastic optimization. arXiv preprint arXiv:1412.6980 (2014)
14. Kortylewski, A., He, J., Liu, Q., Yuille, A.L.: Compositional convolutional neural networks: a deep architecture with innate robustness to partial occlusion. In: Proceedings of the IEEE/CVF Conference on Computer Vision and Pattern Recognition, pp. 8940–8949 (2020)
15. Landman, B., Xu, Z., Igelsias, J., Styner, M., Langerak, T., Klein, A.: Miccai multi-atlas labeling beyond the cranial vault-workshop and challenge. In: Proc. MICCAI Multi-Atlas Labeling Beyond Cranial Vault-Workshop Challenge, vol. 5, p. 12 (2015)
16. Liu, Q., Nickel, M., Kiela, D.: Hyperbolic graph neural networks. Advances in Neural Information Processing Systems 32 (2019)
17. Liu, X., Thermos, S., Sanchez, P., O'Neil, A.Q., Tsaftaris, S.A.: vmfnet: Compositionality meets domain-generalised segmentation. In: International Conference on Medical Image Computing and Computer-Assisted Intervention, pp. 704–714. Springer, Cham (2022). https://doi.org/10.1007/978-3-031-16449-1_67
18. Madry, A., Makelov, A., Schmidt, L., Tsipras, D., Vladu, A.: Towards deep learning models resistant to adversarial attacks. arXiv preprint arXiv:1706.06083 (2017)
19. Milletari, F., Navab, N., Ahmadi, S.A.: V-net: fully convolutional neural networks for volumetric medical image segmentation. In: 2016 Fourth International Conference on 3D Vision (3DV), pp. 565–571. IEEE (2016)
20. Nickel, M., Kiela, D.: Poincaré embeddings for learning hierarchical representations. In: Advances in neural information processing systems 30 (2017)
21. Pérez-García, F., Sparks, R., Ourselin, S.: Torchio: a python library for efficient loading, preprocessing, augmentation and patch-based sampling of medical images in deep learning. Comput. Methods Programs Biomed. **208**, 106236 (2021)
22. Ronneberger, O., Fischer, P., Brox, T.: U-Net: convolutional networks for biomedical image segmentation. In: Navab, N., Hornegger, J., Wells, W.M., Frangi, A.F. (eds.) MICCAI 2015. LNCS, vol. 9351, pp. 234–241. Springer, Cham (2015). https://doi.org/10.1007/978-3-319-24574-4_28
23. Santhirasekaram, A., Kori, A., Winkler, M., Rockall, A., Glocker, B.: Vector quantisation for robust segmentation. In: Wang, L., Dou, Q., Fletcher, P.T., Speidel, S., Li, S. (eds.) MICCAI 2022, pp. 663–672. Springer, Cham (2022). https://doi.org/10.1007/978-3-031-16440-8_63
24. Tramer, F., Boneh, D.: Adversarial training and robustness for multiple perturbations. Advances in Neural Information Processing Systems 32 (2019)
25. Van Den Oord, A., Vinyals, O., et al.: Neural discrete representation learning. Advances in neural information processing systems 30 (2017)
26. Wang, Z., Luo, Y., Qiu, R., Huang, Z., Baktashmotlagh, M.: Learning to diversify for single domain generalization. In: Proceedings of the IEEE/CVF International Conference on Computer Vision, pp. 834–843 (2021)

27. Xu, Z., Liu, D., Yang, J., Raffel, C., Niethammer, M.: Robust and generalizable visual representation learning via random convolutions. arXiv preprint arXiv:2007.13003 (2020)
28. Zhang, H., Cisse, M., Dauphin, Y.N., Lopez-Paz, D.: mixup: beyond empirical risk minimization. arXiv preprint arXiv:1710.09412 (2017)
29. Zhang, H., Rich, P.D., Lee, A.K., Sharpee, T.O.: Hippocampal spatial representations exhibit a hyperbolic geometry that expands with experience. Nature Neuroscience, pp. 1–9 (2022)
30. Zhang, L., Wang, X., Yang, D., Sanford, T., Harmon, S., Turkbey, B., Wood, B.J., Roth, H., Myronenko, A., Xu, D., et al.: Generalizing deep learning for medical image segmentation to unseen domains via deep stacked transformation. IEEE Trans. Med. Imaging **39**(7), 2531–2540 (2020)

Realistic Data Enrichment for Robust Image Segmentation in Histopathology

Sarah Cechnicka[1]([✉]), James Ball[1], Hadrien Reynaud[1], Callum Arthurs[2], Candice Roufosse[2], and Bernhard Kainz[1,3]

[1] Departament of Computing, Imperial College London, London, UK
sc7718@imperial.ac.uk
[2] Centre for Inflammatory Disease, Imperial College London, London, UK
[3] Friedrich-Alexander University Erlangen-Nürnberg, Erlangen, DE, Germany

Abstract. Poor performance of quantitative analysis in histo-pathological Whole Slide Images (WSI) has been a significant obstacle in clinical practice. Annotating large-scale WSIs manually is a demanding and time-consuming task, unlikely to yield the expected results when used for fully supervised learning systems. Rarely observed disease patterns and large differences in object scales are difficult to model through conventional patient intake. Prior methods either fall back to direct disease classification, which only requires learning a few factors per image, or report on average image segmentation performance, which is highly biased towards majority observations. Geometric image augmentation is commonly used to improve robustness for average case predictions and to enrich limited datasets. So far no method provided sampling of a realistic posterior distribution to improve stability, *e.g.* for the segmentation of imbalanced objects within images. Therefore, we propose a new approach, based on diffusion models, which can enrich an imbalanced dataset with plausible examples from underrepresented groups by conditioning on segmentation maps. Our method can simply expand limited clinical datasets making them suitable to train machine learning pipelines, and provides an interpretable and human-controllable way of generating histopathology images that are indistinguishable from real ones to human experts. We validate our findings on two datasets, one from the public domain and one from a Kidney Transplant study. [1](The source code and trained models will be publicly available at the time of the conference, on huggingface and github.)

1 Introduction

Large scale datasets with accurate annotations are key to the successful development and deployment of deep learning algorithms for computer vision tasks. Such datasets are rarely available in medical imaging due to privacy concerns and high cost of expert annotations. This is particularly true for histopathology, where gigapixel images have to be processed [32]. This is one of the reasons

---

S. Cechnicka and J. Ball—Equal contribution.

© The Author(s), under exclusive license to Springer Nature Switzerland AG 2024
L. Koch et al. (Eds.): DART 2023, LNCS 14293, pp. 63–72, 2024.
https://doi.org/10.1007/978-3-031-45857-6_7

why histopathology is, to date, a field in which image-based automated quantitative analysis methods are rare. In radiology, for example, most lesions can be characterised manually into clinically actionable information, *e.g.* measuring the diameter of a tumour. However, this is not possible in histopathology, as quantitative assessment requires thousands of structures to be identified for each case, and most of the derived information is still highly dependent on the expertise of the pathologist. Therefore, supervised Machine Learning (ML) methods quickly became a research focus in the field, leading to the emergence of prominent early methods [24] and, more recently, to high-throughput analysis opportunities for the clinical practice [9,14,22]. Feature location, shape, and size are crucial for diagnosis; this high volume of information required makes automatic segmentation essential for computational pathology [14]. The automated extraction of these features should lead to the transition from their time-consuming and error-prone manual assessment to reproducible quantitative metrics-driven analysis, enabling more robust decision-making. Evaluating biopsies with histopathology continues to be the gold standard for identifying organ transplant rejection [21]. However, imbalances and small training sets still prevent deep learning methods from revolutionizing clinical practice in this field.

In this work, we are interested in the generation of training data for the specific case of histopathology image analysis for kidney transplant biopsies. In order to maximize transplant survival rates and patient well-being, it is essential to identify conditions that can result in graft failure, such as rejection, early on. The current diagnostic classification system presents shortcomings for biopsy assessment, due to its qualitative nature, high observer variability, and lack of granularity in crucial areas [30].

Contribution: We propose a novel data enrichment method using diffusion models conditioned on masks. Our model allows the generation of photo-realistic histopathology images with corresponding annotations to facilitate image segmentation in unbalanced datasets or cases out of distribution. In contrast to conventional geometric image augmentation, we generate images that are indistinguishable from real samples to human experts and provide means to precisely control the generation process through segmentation maps. Our method can also be used for expert training, as it can cover the extreme ends of pathological representations through manual definition of segmentation masks.

Related Work: Diffusion Models have experienced fast-rising popularity [20, 23,25]. Many improvements have been proposed [26,28], some of them suggesting image-to-image transfer methods that can convert simple sketches into photo-realistic images [2]. This is partially related to our approach. However, in contrast to sketch-based synthesis of natural images, we aim at bootstrapping highly performing image segmentation methods from poorly labelled ground truth data.

Data enrichment through synthetic images has been a long-standing idea in the community [8,18,31]. So far, this approach was limited by the generative capabilities of auto-encoding [15] or generative adversarial approaches [6]. A domain gap between real and synthetic images often leads to shortcut learning [5] and biased results with minimal gains. The best results have surprisingly been achieved, not with generative models, but with data imputation by mixing existing training samples to new feature combinations [4,29]. Sample mixing can be combined with generative models like Generative Adversarial Networks (GAN) to enrich the data [18].

2 Method

We want to improve segmentation robustness. We denote the image space as $\mathcal{X}$ and label mask space as $\mathcal{Y}$. Formally, we look for different plausible variations within the joint space $\mathcal{X} \times \mathcal{Y}$ in order to generate extensive datasets $d_k = \{(\boldsymbol{x}_n^{(k)}, \boldsymbol{y}_n^{(k)})\}_{n=1}^{N_k}$, where N_k is the number of labelled data points in the k-th dataset. We hypothesise that training a segmentation network M_θ on combinations of d_k, $d_a \cup d_b \cup \cdots \cup d_c$ with or without samples from an original dataset, will lead to state-of-the-art segmentation performance. We consider any image segmentation model $M_\theta : \mathcal{X} \rightarrow \mathcal{Y}$ that performs pixel-wise classification, $i.e.$ semantic segmentation, in $\mathbb{R}^C$, where C is the number of classes in $\mathcal{Y}$. Thus, predictions for the individual segmentation class labels can be defined as $p(\boldsymbol{y}|\boldsymbol{x}, \theta) = \hat{\boldsymbol{y}} = softmax(M_\theta(\boldsymbol{x}))$.

Inverting the segmentation prediction to $p(\boldsymbol{x}|\boldsymbol{y}, \theta)$ is impractical, as the transformation M_θ is not bijective, and thus inverting it would yield a set of plausible samples from $\mathcal{X}$. However, the inversion can be modelled through a constrained sampling method, yielding single plausible predictions $\hat{\boldsymbol{x}} \in \hat{\mathcal{X}}$ given $\boldsymbol{y} \in \mathcal{Y}$ and additional random inputs $z \sim \mathcal{N}(0, \sigma)$ holding the random state of our generative process. Modelling this approach can be achieved through diffusion probabilistic models [11]. We can thus define $D_\phi : \mathcal{Z} \rightarrow \hat{\mathcal{X}}$ where $\mathcal{Z}$ is a set of Gaussian noise samples. This model can be further conditioned on label masks $\boldsymbol{y}$ and produce matching elements to the joint space $\mathcal{X} \times \mathcal{Y}$ yielding $D_\xi : \mathcal{Z} \times \mathcal{Y} \rightarrow \hat{\mathcal{X}}$.

The first step of our approach, shown in Fig. 1, is to generate a set of images $X_1 = \{\boldsymbol{x}_n^{(1)} | \boldsymbol{x}_n^{(1)} = D_\phi(z), z \sim \mathcal{N}(0, \sigma)\} \subset \hat{\mathcal{X}}$ where D_ϕ is an unconditional diffusion model trained on real data samples. We then map all samples $\boldsymbol{x}_n^{(1)}$ to the corresponding elements in the set of predicted label masks $Y_1 = \{\boldsymbol{y}_n^{(1)} | \boldsymbol{y}_n^{(1)} = M_\theta(\boldsymbol{x}_n^{(1)}), \boldsymbol{x}_n^{(1)} \in X_1\} \subset \hat{\mathcal{Y}}$, where M_θ is a segmentation model trained on real data pairs. This creates a dataset noted d_1. The second step is to generate a dataset d_2, by using a conditional diffusion model D_ξ trained on real images and applied to the data pairs in d_1, such that $X_2 = \{\boldsymbol{x}_n^{(2)} | \boldsymbol{x}_n^{(2)} = D_\xi(\boldsymbol{y}_n^{(1)}, z), \boldsymbol{y}_n^{(1)} \in Y_1, z \sim \mathcal{N}(0, \sigma)\}$. This lets us generate a much larger and more diverse dataset of image-label pairs, where the images are generated from the labels. Our final step is to use this dataset to train a new segmentation model M_ζ that largely outperforms M_θ. To do so, we first train M_ζ on the generated dataset d_2 and fine-tune it on the real dataset.

Image Generation: Diffusion models are a type of generative model producing image samples from Gaussian noise. The idea is to reverse a forward Markovian diffusion process, which gradually adds Gaussian noise to a real image $\boldsymbol{x}_0$ as a time sequence $\{\boldsymbol{x}_t\}_{t=1...T}$. The probability distribution q for the forward sampling process at time t can be written as a function of the original sample image

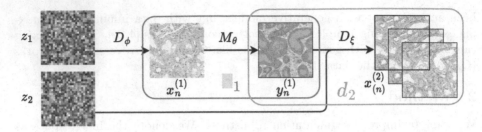

Fig. 1. Summary of our dataset generation approach as described in Sect. 2. We use our diffusion model D_ϕ to generate images, M_θ segments them and D_ξ creates multiple images from these segmentations. Dataset d_2 is the one used to train our final model M_ζ.

$$q\left(\mathbf{x}_t \mid \mathbf{x}_0\right) = \mathcal{N}\left(\mathbf{x}_t; \sqrt{\bar{\alpha}_t}\mathbf{x}_0, (1-\bar{\alpha}_t)\,\mathbf{I}\right), q\left(\mathbf{x}_t \mid \mathbf{x}_s\right) = \mathcal{N}\left(\mathbf{x}_t; (\alpha_t/\alpha_s)\,\mathbf{x}_s, \sigma_{t|s}^2\mathbf{I}\right),$$
(1)

where $\bar{\alpha}_t = \sqrt{1/(1+e^{-\lambda_t})}$ and $\sigma_{t|s}^2 = \sqrt{(1-e^{\lambda_t-\lambda_s})\,\sigma_t^2}$ parameterise the variance of the noise schedule, whose logarithmic signal to noise ratio $\lambda_t = \log\left(\alpha_t^2/\sigma_t^2\right)$ is set to decrease with t [25,27]. The joint distribution p_θ describing the corresponding reverse process is

$$p_\theta\left(\mathbf{x}_{0:T}\right) := p\left(\mathbf{x}_T\right)\prod_{t=1}^{T} p_\theta\left(\mathbf{x}_{t-1} \mid \mathbf{x}_t\right), \quad p_\theta\left(\mathbf{x}_{t-1} \mid \mathbf{x}_t\right) := \mathcal{N}\left(\mathbf{x}_{t-1}; \mu_\theta\left(\mathbf{x}_t, t, \mathbf{c}\right), \sigma_t\right),$$
(2)

where μ_θ is the parameter to be estimated, σ_t is given and $\mathbf{c}$ is an additional conditioning variable. Distribution p depends on the entire dataset and is modelled by a neural network. [11] have shown that learning the variational lower bound on the reverse process is equivalent to learning a model for the noise added to the image at each time step. By modelling $\mathbf{x}_t = \alpha_t\mathbf{x}_0 + \sigma_t\epsilon$ with $\epsilon \sim \mathcal{N}(\mathbf{0}, \mathbf{I})$ we aim to estimate the noise $\epsilon_\theta(x_t, \lambda_t, \mathbf{c})$ in order to minimise the loss function

$$\mathcal{L} = \mathbb{E}_{\epsilon, \lambda_t, \mathbf{c}}\left[w(\lambda_t)\,\|\epsilon_\theta(x_t, \lambda_t, \mathbf{c}) - \epsilon\|^2\right],$$
(3)

where $w(\lambda_t)$ denotes the weight assigned at each time step [12]. We follow [25] using a cosine schedule and DIMM [28] continuous time steps for training and sampling. We further use classifier free guidance [12] avoiding the use of a separate classifier network. The network partly trains using conditional input and partly using only the image such that the resulting noise is a weighted average:

$$\tilde{\epsilon}_\theta\left(x_t, \lambda_t, \mathbf{c}\right) = (1+w)\epsilon_\theta\left(x_t, \lambda_t, \mathbf{c}\right) - w\epsilon_\theta\left(\mathbf{z}_\lambda\right).$$
(4)

The model can further be re-parameterized using v-parameterization [26] by predicting $\mathbf{v} \equiv \alpha_t\epsilon - \sigma_t\mathbf{x}$ rather than just the noise, ϵ, as before. With v-parameterization, the predicted image for time step t is now $\hat{\mathbf{x}} = \alpha_t\mathbf{z}_t - \sigma_t\hat{\mathbf{v}}_\theta(\mathbf{z}_t)$.

Mask Conditioning: Given our proprietary set of histopathology patches, only a small subset of these come with their corresponding segmentation labels.

Therefore, when conditioning on segmentation masks, we first train a set of unconditioned cascaded diffusion models using our unlabelled patches. This allows the model to be pre-trained on a much richer dataset, reducing the amount of labelled data needed to get high-quality segmentation-conditioned samples. Conditioning is achieved by concatenating the segmentation mask, which is empty in pre-training, with the noisy image as input into each diffusion model, at every reverse diffusion step. After pre-training, we fine-tune the cascaded diffusion models on the labelled image patches so that the model learns to associate the labels with the structures it has already learnt to generate in pre-training.

Mask Generation: We use a nnU-Net [13] to generate label masks through multi-class segmentation. The model is trained through a combination of Dice loss $\mathcal{L}_{Dice}$ and Cross-Entropy loss $\mathcal{L}_{CE}$. $\mathcal{L}_{Dice}$ is used in combination with a Cross Entropy Loss $\mathcal{L}_{CE}$ to obtain more gradient information during training [13], by giving it more mobility across the logits of the output vector. Additional auxiliary Dice losses are calculated at lower levels in the model. The total loss function for mask generation can therefore be described with

$$\mathcal{L} = \mathcal{L}_{Dice} + \mathcal{L}_{CE} + \beta(\mathcal{L}_{Dice_{1/2}} + \mathcal{L}_{Dice_{1/4}}), \tag{5}$$

where $\mathcal{L}_{Dice_{1/2}}$ and $\mathcal{L}_{Dice_{1/4}}$ denote the dice auxiliary losses calculated at a half, and a quarter of the final resolution, respectively.

We train two segmentation models M_θ and M_ζ. First, for M_θ, we train the nnU-Net on the original data and ground truth label masks. M_θ is then used to generate the label maps for all the images in d_1, the pool of images generated with our unconditional diffusion model D_ϕ. The second nnU-Net, M_ζ, is pre-trained on our dataset d_2 and we fine-tune it on the original data to produce our final segmentation model.

3   Evaluation

Datasets and Preprocessing: We use two datasets for evaluation. The first one is the public KUMAR dataset [17], which we chose to be able to compare with the state-of-the-art directly. KUMAR consists of 30 WSI training images and 14 test images of 1000×1000 pixels with corresponding labels for tissue and cancer type (`Breast`, `Kidney`, `Liver`, `Prostate`, `Bladder`, `Colon`, and `Stomach`). During training, each raw image is cropped into a patch of 256×256 and then resized to 64×64 pixels. Due to the very limited amount of data available, we apply extensive data augmentation, including rotation, flipping, color shift, random cropping and elastic transformations. However the baseline methods [18] only use 16 of the 30 images available for training.

The second dataset is a proprietary collection of Kidney Transplant Pathology WSI slides with an average resolution of 30000×30000 per slide. These images were tiled into overlapping patches of 1024×1024 pixels. For this work, 1654 patches, classified as kidney cortex, were annotated (glomerulies, tubules, arteries and other vessels) by a consultant transplant pathologist with more than ten years of experience and an assistant with 5 years of experience. Among these, 68 patches, belonging to 6 separate WSI, were selected for testing, while the rest were used for training. The dataset also includes tabular data of patient outcomes and history of creatinine scores before and after the transplant. We resize the 1024×1024 patches down to 64×64 resolution and apply basic shifts, flips and rotations to augment the data before using it to train our first diffusion model. We apply the same transformations but with a higher re-scaling of 256×256 for the first super-resolution diffusion model. The images used to train the second and final super-resolution model are not resized but are still augmented the same. We set most of our training parameters similar to the suggested ones in [25], but use the creatinine scores and patient outcomes as conditioning parameters for our diffusion models.

Implementation: We use a set of three cascaded diffusion models similar to [25], starting with a base model that generates 64×64 images, and then two super-resolution models to upscale to resolutions 256×256 and 1024×1024. Conditioning augmentation is used in super-resolution networks to improve sample quality. In contrast to [25], we use v-parametrization [26] to train our super-resolution models ($64 \times 64 \rightarrow 256 \times 256$ and $256 \times 256 \rightarrow 1024 \times 1024$). These models are much more computationally demanding at each step of the reverse diffusion process, and it is thus crucial to reduce the number of steps during sampling to maintain the sampling time in a reasonable range. We find v-parametrization to allow for as few as 256 steps, instead of 1024 in the noise prediction setting, for the same image quality, while also converging faster. We keep the noise-prediction setting for our base diffusion model, as sampling speed is not an issue at the 64×64 scale, and changing to v-parametrization with 256 time steps generates images with poorer quality in this case. We use PyTorch v1.6 and consolidated [13,25] into the same framework. Three Nvidia A5000 GPUs are used to train and evaluate our models. All diffusion models were trained with over 120,000 steps. The kidney study segmentation models were trained for 200 epochs and fine-tuned for 25, the KUMAR study used 800 epochs and was fine-tuned for 300. Training takes about 10 d and image generation takes 60 s per image. Where real data was used for fine-tuning this was restricted to 30% of the original dataset for kidney images. Diffusion models were trained with a learning rate of 1e−4 and segmentation models were pre-trained with a learning rate of 1e−3 which dropped to 3e−6 when no change was observed on the validation set in 15 epochs. Through D_ϕ, M_θ and D_ξ the number of synthetic samples matched the number of real ones. All models used Adam optimiser. See the supplemental material for further details about the exact training configurations.

Setup: We evaluate the performance of nnU-Net [13] trained on the data enriched by our method. We trained over 5 different combinations of training sets, using the same test set for metrics comparison, and show the results in Table 1. First, we train a base nnU-Net solely on real image data, (1), before fine-tuning it, independently, twice: once with a mixture of real and synthetic images as (2), and once exclusively with synthetic images as (3). The 4th and 5th models correspond to nnU-Nets retrained from scratch using exclusively synthetic images as (4), and one further fine-tuned on real images as (5) in Table 1.

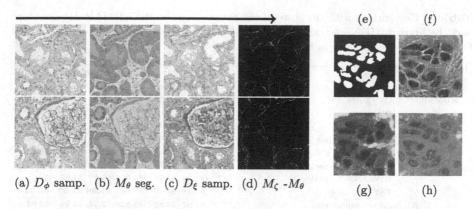

(a) D_ϕ samp.   (b) M_θ seg.   (c) D_ξ samp.   (d) M_ζ -M_θ

Fig. 2. Left: Outputs from our models. From left to right: (a) sample from D_ϕ, (b) overlaid segmentation from M_θ, (c) sample from D_ξ, (d) difference map of segmentation from M_ζ and M_θ highlighting shape improvement. Segmentation colors are: red: Tubuli, blue: Glomeruli, green: Vessels. Right: Diffusion model-generated images conditioned on different tissue types in KUMAR, using the same label mask (e). Generated images are from Liver (f), Bladder (g), and Breast (h) tissues. This shows that our conditioning seems to allow a plausible mask to produce any kind of tissue.

Results: Our quantitative results are summarised and compared to the state-of-the-art in Table 1 using the Dice coefficient (Dice) and Aggregated Jaccard Index (AJI) as suggested by [18]. Qualitative examples are provided in Fig. 2 (left), which illustrates that our model can convert a given label mask into a number of different tissue types and Fig. 2, where we compare synthetic enrichment images of various tissue types from our kidney transplant data.

Sensitivity Analysis: Out of our 5 models relying on additional synthetic data in the KUMAR dataset experiments, all outperform previous SOTA on the Dice score. Importantly, synthetic results allow for high performance in previously unseen tissue types. Results are more nuanced when it comes to the AJI, as AJI over-penalizes overlapping regions [7]. Additionally, while a further AJI loss was introduced to the final network M_ζ, loss reduction, early stopping and the M_θ networks do not take it into account. Furthermore, Table 1 shows that, for the KIDNEY dataset, we can reach high performance (88% Dice) while training M_ζ on 30% (500 samples) of the real KIDNEY data (1). We also observe that the model pretrained on synthetic data and fine-tuned on 500 real images (5), outperforms the one only trained on 500 real images (1). Additionally, we discover that training the model on real data before fine-tuning it on synthetic samples (3) does not work as well as the opposite approach. We argue that pre-training an ML model on generated data gives it a strong prior on large dataset distributions and alleviates the need for many real samples in order to learn the final, exact, decision boundaries, making the learning procedure more data efficient.

Discussion: We have shown that data enrichment with generative diffusion models can help to boost performance in low data regimes, *e.g.*, KUMAR data,

Table 1. Comparison with the state-of-the-art methods on the KUMAR dataset (top) and the limited KIDNEY transplant dataset (bottom). Metrics are chosen as in [18]: Dice and AJI. Best values in bold.

	Method	Dice (%) Seen	Unseen	All	AJI (%) Seen	Unseen	All
KUMAR	CNN3 [16]	82.26	83.22	82.67	51.54	49.89	50.83
	DIST [19]	-	-	-	55.91	56.01	55.95
	NB-Net [3]	79.88	80.24	80.03	59.25	53.68	56.86
	Mask R-CNN [10]	81.07	82.91	81.86	59.78	55.31	57.86
	HoVer-Net [7] (*Res50)	80.60	80.41	80.52	59.35	56.27	58.03
	TAFE [1] (*Dense121)	80.81	83.72	82.06	61.51	61.54	61.52
	HoVer-Net + InsMix [18]	80.33	81.93	81.02	59.40	57.67	58.66
	TAFE + InsMix [18]	81.18	84.40	82.56	**61.98**	**65.07**	**63.31**
Ours	(1) trained on real	82.97	84.89	83.52	52.34	54.29	52.90
	(2) fine-tuned by synthetic+real	**87.82**	88.66	**88.06**	60.79	60.05	60.71
	(3) fine-tuned by synthetic	87.12	87.52	87.24	59.53	58.85	59.33
	(4) trained on synthetic	86.06	**89.69**	87.10	52.89	58.93	54.62
	(5) trained on synthetic, fine-tuned on real	85.75	87.88	86.36	56.01	57.83	56.5
KIDNEY Ours	(1) trained on real (30% data)			88.01			62.05
	(2) fine-tuned by synthetic+real			92.25			69.11
	(3) fine-tuned by synthetic			89.65			58.59
	(4) trained on synthetic			82.00			42.40
	(5) trained on synthetic, fine-tuned on real			**92.74**			**71.55**

but also observe that when using a larger dataset, where maximum performance might have already been reached, the domain gap may become prevalent and no further improvement can be observed, *e.g.*, full KIDNEY data (94% Dice). Estimating the upper bound for the required labelled ground truth data for image segmentation is difficult in general. However, testing model performance saturation with synthetic data enrichment might be an experimental way forward to test for convergence bounds. Finally, the best method for data enrichment seems to depend on the quality of synthetic images.

4   Conclusion

In this paper, we propose and evaluate a new data enrichment and image augmentation scheme based on diffusion models. We generate new, synthetic, high-fidelity images from noise, conditioned on arbitrary segmentation masks. This allows us to synthesise an infinite amount of plausible variations for any given feature arrangement. We have shown that using such enrichment can have a drastic effect on the performance of segmentation models trained from small datasets used for histopathology image analysis, thus providing a mitigation strategy for expensive, expert-driven, manual labelling commitments.

Acknowledgements. This work was supported by the UKRI Centre for Doctoral Training in Artificial Intelligence for Healthcare (EP/S023283/1). Dr. Roufosse is supported by the National Institute for Health Research (NIHR) Biomedical Research Centre based at Imperial College Healthcare NHS Trust and Imperial College London. The views expressed are those of the authors and not necessarily those of the NHS, the NIHR or the Department of Health. Dr Roufosse's research activity is made possible with generous support from Sidharth and Indira Burman. The authors gratefully acknowledge the scientific support and HPC resources provided by the Erlangen National High Performance Computing Center (NHR@FAU) of the Friedrich-Alexander-Universität Erlangen-Nürnberg (FAU) under the NHR projects b143dc and b180dc. NHR funding is provided by federal and Bavarian state authorities. NHR@FAU hardware is partially funded by the German Research Foundation (DFG) - 440719683. Additional support was also received by the ERC - project MIA-NORMAL 101083647, DFG KA 5801/2-1, INST 90/1351-1 and by the state of Bavarian.

References

1. Chen, S., Ding, C., Tao, D.: Boundary-assisted region proposal networks for nucleus segmentation. In: Martel, A.L., et al. (eds.) MICCAI 2020. LNCS, vol. 12265, pp. 279–288. Springer, Cham (2020). https://doi.org/10.1007/978-3-030-59722-1_27
2. Cheng, S.I., Chen, Y.J., Chiu, W.C., Tseng, H.Y., Lee, H.Y.: Adaptively-realistic image generation from stroke and sketch with diffusion model. In: CVPR 2023, pp. 4054–4062 (2023)
3. Cui, Y., Zhang, G., Liu, Z., Xiong, Z., Hu, J.: A deep learning algorithm for one-step contour aware nuclei segmentation of histopathology images. Med. Biol. Eng. Comput. **57**, 2027–2043 (2019)
4. Dwibedi, D., Misra, I., Hebert, M.: Cut, paste and learn: surprisingly easy synthesis for instance detection. In: IEEE ICCV 2017, pp. 1301–1310 (2017)
5. Geirhos, R., et al.: Shortcut learning in deep neural networks. Nat. Mach. Intell. **2**(11), 665–673 (2020)
6. Goodfellow, I., et al.: Generative adversarial networks. Commun. ACM **63**(11), 139–144 (2020)
7. Graham, S., et al.: Hover-net: simultaneous segmentation and classification of nuclei in multi-tissue histology images. Med. Image Anal. **58**, 101563 (2019)
8. Gupta, L., Klinkhammer, B.M., Boor, P., Merhof, D., Gadermayr, M.: GAN-based image enrichment in digital pathology boosts segmentation accuracy. In: Shen, D., et al. (eds.) MICCAI 2019. LNCS, vol. 11764, pp. 631–639. Springer, Cham (2019). https://doi.org/10.1007/978-3-030-32239-7_70
9. Han, Z., Wei, B., Zheng, Y., Yin, Y., Li, K., Li, S.: Breast cancer multi-classification from histopathological images with structured deep learning model. Sci. Rep. **7**(1), 4172 (2017)
10. He, K., Gkioxari, G., Dollár, P., Girshick, R.: Mask r-cnn. In: IEEE ICCV 2017, pp. 2961–2969 (2017)
11. Ho, J., Jain, A., Abbeel, P.: Denoising diffusion probabilistic models. Adv. Neural. Inf. Process. Syst. **33**, 6840–6851 (2020)
12. Ho, J., Salimans, T.: Classifier-free diffusion guidance. arXiv preprint arXiv:2207.12598 (2022)
13. Isensee, F., Jaeger, P.F.: nnU-Net: a self-configuring method for deep learning-based biomedical image segmentation. Nat. Methods. https://doi.org/10.1038/s41592-020-01008-z
14. Khened, M., Kori, A., Rajkumar, H., Krishnamurthi, G., Srinivasan, B.: A generalized deep learning framework for whole-slide image segmentation and analysis. Sci. Rep. **11**(1), 1–14 (2021)

15. Kingma, D.P., Welling, M.: Auto-encoding variational bayes. arXiv preprint arXiv:1312.6114 (2013)
16. Kumar, N., et al.: A multi-organ nucleus segmentation challenge. IEEE Trans. Med. Imaging **39**(5), 1380–1391 (2019)
17. Kumar, N., Verma, R., Sharma, S., Bhargava, S., Vahadane, A., Sethi, A.: A dataset and a technique for generalized nuclear segmentation for computational pathology. IEEE Trans. Med. Imaging **36**(7), 1550–1560 (2017). https://doi.org/10.1109/TMI.2017.2677499
18. Lin, Y., Wang, Z., Cheng, K.T., Chen, H.: Insmix: towards realistic generative data augmentation for nuclei instance segmentation. In: MICCAI 2022, Part II, pp. 140–149. Springer (2022). https://doi.org/10.1007/978-3-031-16434-7_14
19. Naylor, P., Laé, M., Reyal, F., Walter, T.: Segmentation of nuclei in histopathology images by deep regression of the distance map. IEEE Trans. Med. Imaging **38**(2), 448–459 (2018)
20. Ramesh, A., Pavlov, M., Goh, G., Gray, S., Voss, C., Radford, A.: Zero-Shot Text-to-Image Generation. arXiv:2102.12092 (February 2021)
21. Reeve, J., et al.: Diagnosing rejection in renal transplants: a comparison of molecular-and histopathology-based approaches. Am. J. Transplant. **9**(8), 1802–1810 (2009)
22. Reisenbüchler, D., Wagner, S.J., Boxberg, M., Peng, T.: Local attention graph-based transformer for multi-target genetic alteration prediction. In: MICCAI 2022, Part II, pp. 377–386. Springer (2022). https://doi.org/10.1007/978-3-031-16434-7_37
23. Rombach, R., Blattmann, A., Lorenz, D., Esser, P., Ommer, B.: High-Resolution Image Synthesis with Latent Diffusion Models, arXiv:2112.10752 (April 2022)
24. Ronneberger, O., Fischer, P., Brox, T.: U-Net: convolutional networks for biomedical image segmentation. In: Navab, N., Hornegger, J., Wells, W.M., Frangi, A.F. (eds.) MICCAI 2015. LNCS, vol. 9351, pp. 234–241. Springer, Cham (2015). https://doi.org/10.1007/978-3-319-24574-4_28
25. Saharia, C., et al.: Photorealistic text-to-image diffusion models with deep language understanding. arXiv preprint arXiv:2205.11487 (2022)
26. Salimans, T., Ho, J.: Progressive Distillation for Fast Sampling of Diffusion Models (June 2022). arXiv:2202.00512
27. Sohl-Dickstein, J., Weiss, E., Maheswaranathan, N., Ganguli, S.: Deep unsupervised learning using nonequilibrium thermodynamics. In: International Conference on Machine Learning, pp. 2256–2265. PMLR (2015)
28. Song, J., Meng, C., Ermon, S.: Denoising Diffusion Implicit Models, arXiv:2010.02502 (October 2022)
29. Tan, J., Hou, B., Day, T., Simpson, J., Rueckert, D., Kainz, B.: Detecting outliers with poisson image interpolation. In: de Bruijne, M., et al. (eds.) MICCAI 2021. LNCS, vol. 12905, pp. 581–591. Springer, Cham (2021). https://doi.org/10.1007/978-3-030-87240-3_56
30. van Loon, E., et al.: Forecasting of patient-specific kidney transplant function with a sequence-to-sequence deep learning model. JAMA Netw. Open **4**(12), e2141617–e2141617 (2021)
31. Wang, J., Perez, L., et al.: The effectiveness of data augmentation in image classification using deep learning. Convolut. Neural Netw. Vis. Recognit. **11**(2017), 1–8 (2017)
32. Ye, J., et al.: Synthetic sample selection via reinforcement learning. In: Martel, A.L., et al. (eds.) MICCAI 2020. LNCS, vol. 12261, pp. 53–63. Springer, Cham (2020). https://doi.org/10.1007/978-3-030-59710-8_6

Boosting Knowledge Distillation via Random Fourier Features for Prostate Cancer Grading in Histopathology Images

Trinh Thi Le Vuong and Jin Tae Kwak[✉]

School of Electrical Engineering, Korea University, Seoul, Republic of Korea
jkwak@korea.ac.kr

Abstract. There has been a growing number of pathology image datasets, in particular for cancer diagnosis. Although these datasets permit easy access and development of computational pathology tools, the current computational models still struggle to handle unseen datasets due to various reasons. Transfer learning and fine-tuning are standard techniques to adapt an existing model that was trained on one dataset to another. However, this approach does not fully exploit the existing model and the target dataset. Inspired by knowledge distillation, we propose a student-teacher strategy that distills knowledge from a well-trained teacher model, generally trained on a larger dataset, to a student model to be tested on a small dataset. To facilitate efficient and effective knowledge distillation and transfer, we employ contrastive learning and non-parameterized random Fourier features for compressed feature mapping into a lower-dimensional space. We evaluated our proposed method using three prostate cancer datasets, including a teacher dataset, a target student dataset, and an independent test dataset. The experimental results demonstrate that the proposed approach outperforms other transfer learning and state-of-the-art knowledge distillation methods. Code is available at: https://github.com/trinhvg/KD_CoRFF.

Keywords: knowledge distillation · random Fourier features · contrastive learning · cancer grading · digital pathology

1 Introduction

Prostate cancer is major cancer affecting men worldwide [29]. The definitive diagnosis of prostate cancer is, by and large, relying on manual examination of tissue specimens by human experts, which is slow and subjective. With the increasing burden of prostate cancer [8], there is a need for developing a fast and reliable method to improve the accuracy and efficiency of cancer diagnosis. Recently, computational pathology, which conducts automated image analysis and interpretation mainly based upon deep learning techniques, has shown great potential for predicting and grading various types of cancers in pathology images. However, computational pathology tools still suffer from the shortage of high-quality

© The Author(s), under exclusive license to Springer Nature Switzerland AG 2024
L. Koch et al. (Eds.): DART 2023, LNCS 14293, pp. 73–83, 2024.
https://doi.org/10.1007/978-3-031-45857-6_8

and diverse datasets, and thus it is susceptible to domain-shift problems [27]. Although there is a growing number of publicly available datasets in computational pathology [2], when it comes to a specific task, researchers are unlikely to have access to sufficient datasets and appropriate ground truth labels.

To resolve or alleviate such problems, most of the current works employ a transfer learning approach; for instance, transferring the pre-trained weights from ImageNet [15] to a target model or domain. Although it has been successful, computational pathology tools may benefit more from the weights or models that were trained on the same or relevant domain. Knowledge distillation (KD) is an approach to transferring knowledge from one model, designated as a teacher model, to another model, so-called a student model. It is a simple but useful method that can utilize the existing models. Thus, KD can be used to build an accurate and robust computational pathology tool by exploiting the existing models that were trained on relevant computational pathology datasets. Previous KD methods, however, generally sought to obtain a smaller model that could perform as good as the larger model (teacher model).

Random Fourier features (RFFs) have been heavily studied in the theory of neural networks [13], primarily with respect to kernel approximation [30] and complexity reduction in time and space [16]. Several successful applications of RFFs are to speed up kernel ridge regression [1], to approximate soft-max [23], to accelerate transformers by estimating the softmax function [7,21] and mixing tokens with Fourier transforms [11]. RFFs have also been used for generative models [14,17,26], and low-dimensional data projection in kernel clustering [6]. Although successful in several applications, to the best of our knowledge, it has not been explored in pathology image analysis.

Herein, we develop a robust KD framework for prostate cancer grading in pathology images. The proposed method employs RFFs to compress the high-dimensional feature representations of pathology images from a teacher model and uses the compressed feature representations to train a student model using contrastive learning. Three publicly available prostate cancer datasets are used for evaluation. We distill knowledge from the teacher model trained on the largest dataset and use it to train and test the student model on the rest of the two smaller datasets. The experimental results show that our method outperforms other methods, providing a robust and accurate model for prostate cancer grading on small target datasets.

2   Methods

The proposed method, named CoRFF, aims at building an accurate and robust model for cancer grading on a small target dataset. During training, it learns the knowledge from the existing model that was trained on a large dataset in advance. The proposed method is built based upon three key components: 1) KD, 2) RFFs, and 3) a contrastive loss function. KD provides an overall framework to transfer the knowledge from the existing model to a target model. RFFs and the contrastive loss function permit an efficient and effective knowledge transfer between the two models. A fully connected layer is added on top of the student encoder to conduct the cancer grading task. We freeze the teacher encoder and then train contrastive learning and classification tasks together.

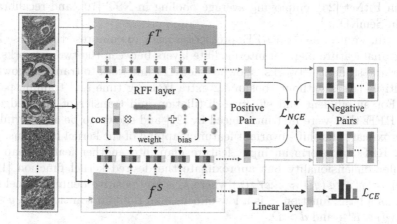

Fig. 1. The overview of the proposed method using a knowledge distillation framework with contrastive loss and RFF layer. Contrastive loss and RFF layer are measured for 5 consecutive feature maps within the network.

2.1 Knowledge Distillation

Suppose that we have a teacher network f^T and a student network f^S. These two networks are built based upon EfficientNetB0 [25], including a sequence of one convolutional layer, 16 mobile inverted bottleneck convolution blocks (MBConv), and one fully connected layer. We split the network into L consecutive blocks by the number of down-sampling layers. Given an input image x, we denote each set of L feature maps of these two models as $\{f_l^S(x)|l = 1, ..., L\}$ and $\{f_l^T(x)|l = 1, ..., L\}$.

The objective of the initial version of KD [9] is to minimize the KL divergence between the last feature maps or outputs that are produced by the teacher model and student model as follows $\mathcal{L}_{KD} = \sum_x f_L^T(x) \log(\frac{f_L^T(x)}{f_L^S(x)})$. Other intermediate feature maps from both models can be incorporated in the same manner. This naïve strategy of matching the corresponding feature maps may be suitable for obtaining a compact model.

2.2 Random Fourier Features (RFFs) Projection Layer

Since most of the current deep neural networks produce high-dimensional feature maps in the latent space (>1 million), such high-dimensional feature maps are likely to be sparse. Hence, the naïve strategy may hinder the learning process in both speed and accuracy. When transferring intermediate-layer features, the high-dimensional feature maps tend to be larger and less semantically informative compared to deeper layers. To address this, various projection layers have been used for compression purposes. Examples include directly matching feature maps in FitNet [24], employing average pooling in NST [10], and recalibrating them in SemKD [5].

Herein, we proposed an RFFs projection layer to compress the sparse high-dimensional feature maps of intermediate layers between the two networks, i.e., teacher and student networks, allowing the KD framework to transfer knowledge at multiple layers without consuming extra training time and training parameters. For an efficient and effective distillation and transfer of knowledge, we adopt RFFs that were first introduced by [22] and shown to be a powerful tool to approximate the shift-invariant feature mapping of the kernel functions. Intuitively, RFFs transform an input feature map into another feature map with a smaller dimensionality but approximate the desired kernel function [12] as $\mathcal{K}(x, y) = \langle \varphi(x), \varphi(y) \rangle \approx z(x)'z(y)$ where $\mathcal{K}$ is any positive definite kernel function, φ is a basis function, z is a low-dimensional random transform, $x, y \in R^d$, $z(x), z(y) \in R^D$, and $d > D$.

In our context, we adopt RFFs to approximate the basis/mapping functions φ for the high-dimensional, intermediate feature maps between the teacher and student networks. We convert the input feature maps into a low-dimensional feature space by the random transform z where the two feature maps from the teacher and student networks can be better utilized for KD. Given an input feature map $f_l \in R^d$, we construct the random transform z by drawing D independently and identically distributed (iid) samples $w_1, ..., w_D \in R^d$ from Gaussian distribution (weights) and D iid samples $b_1, ..., b_D \in R$ from the uniform distribution on $[0, 2\pi]$ (biases). Using the weights $w_1, ..., w_D$ and biases $b_1, ..., b_D$, the random transform z converts the input feature map f_l into a randomized feature map $\psi_l \in R^d$ as follows:

$$\psi_l = \sqrt{\frac{1}{D}} [cos(w_1^T f_l + b_1), cos(w_2^T f_l + b_2), ..., cos(w_D^T f_l + b_D)]^T. \tag{1}$$

The corresponding feature maps in the teacher and student models are converted by the same RFF projection layer (Fig. 1).

2.3   Objective Function

To optimize the proposed method, we employ two loss functions given by $\mathcal{L} = \mathcal{L}_{CE} + \alpha \mathcal{L}_{NCE}$ where $\mathcal{L}_{CE}$ is cross−entropy loss, $\mathcal{L}_{NCE}$ is contrastive loss and α is a hyper-parameter ($\alpha = 1e - 4$). $\mathcal{L}_{CE}$ is used to measure the total entropy between the prediction y^c and ground truth y as follows:

$$\mathcal{L}_{CE}(y, y^c) = -\frac{1}{N} \sum_{i=1}^{N} y_i \log(y_i^c) \tag{2}$$

$\mathcal{L}_{NCE}$ is an extended version of InfoNCE loss [18]. Given a batch of B samples $\{x_i\}_{i=1}^B$, $\mathcal{L}_{NCE}$ maximizes and minimizes the similarity between positive and negative pairs in the feature space, respectively. For an input sample x_i, the feature representation of a positive pair and negative pairs are determined as $\{f_{L,i}^S, f_{L,i}^T\}$ and $\{f_{L,j}^S, f_{L,k}^T\}_{j,k=1,j\neq k}^B$, respectively, that are encoded by the student and teacher model. $\mathcal{L}_{NCE}$ for an input sample x is defined as:

$$\mathcal{L}_{NCE}(f_L^S, f_L^T) = -\mathbb{E}\left[\log \frac{\exp(f_{L,i}^S \cdot f_{L,i}^T / \tau)}{\sum_{j,k=1,j\neq k}^B \exp(f_{L,j}^S \cdot f_{L,k}^T / \tau)}\right] \quad (3)$$

where the temperature hyper-parameter τ is set to 0.07.

Table 1. Results of cancer grading on D_{TestI}

Model	KD	Acc (%)	Recall	Precision	F1	κ_w
TC	no	63.3	0.603	0.542	0.525	0.532
ST_{no}	no	66.7	0.586	0.584	0.584	0.555
$ST_{ImageNet}$	no	67.2	0.624	0.619	0.602	0.565
ST_{PANDA}	no	69.2	0.633	0.624	0.628	0.607
KD_{plain} [9]	yes	66.9	0.590	0.574	0.543	0.547
FitNet [24]	yes	68.6	**0.717**	0.589	0.623	0.602
PKT [20]	yes	71.2	0.668	0.642	0.653	0.646
RKD [19]	yes	69.8	0.667	0.641	0.652	0.630
NST [10]	yes	70.8	0.633	0.679	0.654	0.639
SemCKD [5]	yes	66.3	0.661	0.563	0.580	0.578
SimKD [4]	yes	66.2	0.518	0.602	0.508	0.482
CoRFF (Ours)	yes	**73.2**	0.665	**0.706**	**0.683**	**0.666**

3   Experiment

3.1   Dataset

Three prostate cancer datasets with four pathology classes, such as benign (BN), grade 3 (G3), grade 4 (G4), and grade 5 (G5), are used in this study. The first dataset [3] consists of 5158 whole slide images (WSIs) obtained from the Prostate cANcer graDe Assessment (PANDA) Challenge and were digitized at 20x magnification using a 3DHistech Pannoramic Flash II 250 scanner (0.24 μm/pixel). This dataset is used to train the teacher network. The second dataset, obtained from the Harvard dataverse, includes 886 tissue core images scanned at 40x magnification using a NanoZoomer-XR Digital slide scanner (0.23 μm/pixel). It is split into a training set (D_{Train}) consisting of 2742 BN, 7226 G3, 5114 G4, and

Table 2. Silhouette coefficient scores for D_{TestI}: Evaluating the degree of class feature cluster separation encoded by CoRFF and competing models. A higher silhouette score indicates better cluster separation.

Method	KD	Avg	BN-G3	BN-G4	BN-G5	G3-G4	G4-G5	G4-G5
TC	no	0.253	0.189	0.189	0.391	0.154	0.430	0.162
ST_{no}	no	0.248	0.117	0.247	0.473	0.172	0.390	0.087
$ST_{ImageNet}$	no	0.285	0.252	0.277	0.428	0.170	0.440	0.142
ST_{PANDA}	no	0.293	0.213	0.280	0.455	0.175	0.466	0.170
KD_{plain} [9]	yes	0.247	0.163	0.167	0.405	0.171	0.441	0.138
FitNet [24]	yes	0.255	0.203	0.189	0.362	0.174	0.446	0.158
PKT [20]	yes	0.240	0.172	0.146	0.375	0.169	0.437	0.139
RKD [19]	yes	0.257	0.218	0.164	0.377	0.169	0.457	0.158
NST [10]	yes	0.303	0.235	0.282	0.497	0.166	0.463	0.173
SemCKD [5]	yes	0.114	0.137	0.061	0.082	0.007	0.244	0.153
SimKD [4]	yes	0.235	0.171	0.099	0.372	0.155	0.444	0.170
CoRFF (Ours)	yes	**0.339**	**0.281**	**0.349**	**0.492**	**0.198**	**0.504**	**0.209**

2703 G5 image patches, and a test set (D_{TestI}) comprising 127 BN, 1602 G3, 2121 G4, and 351 G5 image patches. The third dataset (D_{TestII}), acquired from Asan Medical Center [28], contains 20,844 image patches (2381 BN, 10815 G3, 7504 G4, and 144 G5) that were digitized at 40x magnification using an Aperio digital slide scanner (0.2465 µm/pixel). D_{Train}, D_{TestI}, and D_{TestII} have image patches of size 750×750 pixels that are resized to 512×512 pixels.

3.2 Training Procedure

The proposed method is trained with Adam optimizer with default parameters. We trained all the models for 60 epochs, with cosine annealing warm restarts schedule with an initial learning rate of $1.0e - 3$, and restarts after 20 epochs.

The teacher network f^T has been trained on an extensive dataset, while the student network f^S is being trained on a specific target dataset. Notably, we are restricted to utilizing solely the weight configuration of the teacher network f^T, with no access to the actual teacher dataset during the student's training process. This signifies that the teacher network's parameters are fixed, while only the weight associated with the student network f^S undergoes updates throughout its training on the targeted dataset.

Once the student network's training is complete, it is directly applied to both the test set (D_{TestI}) and an independent test set (D_{TestII}) without undergoing any further adaptation or adjustments. The same training strategy is also applied to all comparative experiments.

3.3   Comparative Experiments

We compare the proposed approach with several competing methods: 1) TC: a teacher model f^T only trained on PANDA without fine tuning on D_{Train}, 2) ST_{no}: a student f^S without pre-trained weights, 3) $ST_{ImageNet}$: a student f^S with pre-trained weights on ImageNet, 4) ST_{PANDA}: a student f^S with pre-trained weight on PANDA, 5) KD_{plain} [9], 6) FitNet [24], 7) PKT [20] and 8) RKD [19] 9) NST [10], 10) SemCKD [5] and 11) SimKD [4]. Moreover, we conduct two ablation experiments to investigate the effect of RFFs: 1) KD-NCE: f^S with contrastive loss and 2) KD-NCE-Linear: f^S with a learnable linear layer instead of the RFF layer.

4   Results and Discussions

Table 1 shows the prostate cancer classification results of the proposed method (CoRFF) and other competing methods on D_{TestI}. All the models were trained on D_{Train}. To evaluate the classification results, we adopted five evaluation metrics, including accuracy (Acc), Recall, Precision, F1, and quadratic weighted kappa (κ_w). The experimental results show that CoRFF outperforms other competing methods on prostate cancer grading except for FitNet for recall. Moreover, we have made several observations. First, TC (f^T) was inferior to all other models, i.e., without being exposed to the target dataset, the performance of the models is not guaranteed. Second, the models that were trained using KD were, by and large, superior to the ones without KD, suggesting the advantages of exploiting the teacher model. Third, KD_{plain}, SemCKD and SImKD were inferior to $ST_{ImageNet}$ and ST_{PANDA}, even though these models use both the teacher model and the target dataset. This emphasizes the importance of the effectiveness of knowledge transfer.

Table 2 shows the silhouette coefficient score that measures the degree of separation of class clusters encoded by CoRFF and competitors. The higher the silhouette score, the better separated the clusters are. CoRFF achieved not only the best-averaged silhouette coefficient score but also the highest score for any pairs of class labels, indicating that CoRFF better separates the samples of different class labels, leading to accurate classification results.

Moreover, we applied the models that were trained on D_{Train} to D_{TestII}, which was obtained from a different institute using a different digital scanner. Table 3 shows the performance of CoRFF and other competitive models. Due to the difference between the two datasets, i.e., a domain shift problem, the performance of these methods, in general, degraded in comparison to the results on D_{TestI}. However, similar observations were made on D_{TestII}. CoRFF achieved the best classification performance on three evaluation metrics and the second-best performance for the rest of the metrics. $ST_{ImageNet}$, in particular, obtained the best accuracy of 74.2% but low recall, precision, and F1. In a close examination of the classification results (Fig. 2), we found that $ST_{ImageNet}$ misclassifies most G5 samples, which is critical in clinical applications.

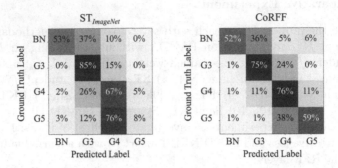

Fig. 2. Confusion matrices of $ST_{ImageNet}$ and CoRFF on D_{TestII}

Table 3. Results of cancer grading on D_{TestII}.

Model	KD	Acc (%)	Recall	Precision	F1	κ_w
TC	no	61.3	0.594	0.603	0.509	0.459
ST_{no}	no	60.6	0.468	0.551	0.373	0.353
$ST_{ImageNet}$	no	**74.2**	0.532	0.596	0.550	0.617
ST_{PANDA}	no	66.8	0.644	0.562	0.532	0.600
KD_{plain} [9]	yes	68.6	0.612	**0.639**	0.569	0.517
FitNet [24]	yes	67.9	0.652	0.537	0.525	0.608
PKT [20]	yes	67.9	0.660	0.576	0.528	0.580
RKD [19]	yes	67.5	0.650	0.583	0.520	0.607
NST [10]	yes	62.7	0.579	0.583	0.469	0.476
SemCKD [5]	yes	59.2	0.533	0.557	0.434	0.409
SimKD [4]	yes	59.3	0.514	0.635	0.49	0.399
CoRFF (Ours)	yes	72.7	**0.732**	0.611	**0.571**	**0.617**

Table 4. Ablation study on the effect of RFFs.

Dataset	Model	Acc (%)	Recall	Precision	F1	κ_w
D_{TestI}	KD-NCE	72.6	0.633	0.692	0.657	0.659
	KD-NCE-linear	71.4	**0.673**	0.658	0.665	0.638
	CoRFF (Ours)	**73.2**	0.665	**0.706**	**0.683**	**0.666**
D_{TestII}	KD-NCE	66.8	0.620	0.585	0.536	0.585
	KD-NCE-Linear	66.7	0.583	0.587	0.502	0.515
	CoRFF (Ours)	**72.7**	**0.732**	**0.611**	**0.571**	**0.617**

Ablation Study: To investigate the effect of RFFs, we conducted two ablation experiments (Table 4). For both datasets, there was a performance drop for the model without RFFs. As RFFs are replaced by a learnable linear layer, the performance was more or less the same as the one without RFFs. It is noteworthy that the performance gain by CoRFF was greater on D_{TestII}, suggesting that RFFs aid in improving the robustness and domain invariance of the model.

5   Conclusion

This study proposes an efficient and effective KD framework for cancer grading in pathology images. Taking advantage of Random Fourier features and contrastive learning, the proposed method is able to train a classification model on a relatively small dataset in an accurate and reliable fashion, which performs well on an independent dataset from a different source. The proposed method is generic and computationally inexpensive. In a follow-up study, we will further improve the proposed method and apply it to other types of cancers and organs.

Acknowledgments. This work was supported by the National Research Foundation of Korea (NRF) (No. 2021R1A2C2014557).

References

1. Avron, H., Clarkson, K.L., Woodruff, D.P.: Faster kernel ridge regression using sketching and preconditioning. SIAM J. Matrix Anal. Appl. **38**(4), 1116–1138 (2017)
2. Bankhead, P.: Developing image analysis methods for digital pathology. J. Pathol. **257**(4), 391–402 (2022)
3. Bulten, W., et al.: Artificial intelligence for diagnosis and Gleason grading of prostate cancer: the panda challenge. Nat. Med. **28**(1), 154–163 (2022)
4. Chen, D., Mei, J.P., Zhang, H., Wang, C., Feng, Y., Chen, C.: Knowledge distillation with the reused teacher classifier. In: Proceedings of the IEEE/CVF Conference on Computer Vision and Pattern Recognition, pp. 11933–11942 (2022)
5. Chen, D., et al.: Cross-layer distillation with semantic calibration. In: Proceedings of the AAAI Conference on Artificial Intelligence, vol. 35, pp. 7028–7036 (2021)
6. Chitta, R., Jin, R., Jain, A.K.: Efficient kernel clustering using random Fourier features. In: 2012 IEEE 12th International Conference on Data Mining, pp. 161–170. IEEE (2012)
7. Choromanski, K.M., et al.: Rethinking attention with performers. In: International Conference on Learning Representations (2021)
8. Culp, M.B., Soerjomataram, I., Efstathiou, J.A., Bray, F., Jemal, A.: Recent global patterns in prostate cancer incidence and mortality rates. Eur. Urol. **77**(1), 38–52 (2020)
9. Hinton, G., Vinyals, O., Dean, J.: Distilling the knowledge in a neural network. arXiv preprint arXiv:1503.02531 (2015)
10. Huang, Z., Wang, N.: Like what you like: knowledge distill via neuron selectivity transfer. arXiv preprint arXiv:1707.01219 (2017)

11. Lee-Thorp, J., Ainslie, J., Eckstein, I., Ontanon, S.: FNet: mixing tokens with Fourier transforms. In: Proceedings of the 2022 Conference of the North American Chapter of the Association for Computational Linguistics: Human Language Technologies, pp. 4296–4313. Association for Computational Linguistics, Seattle (2022)
12. Li, Z., Ton, J.F., Oglic, D., Sejdinovic, D.: Towards a unified analysis of random Fourier features. In: International Conference on Machine Learning, pp. 3905–3914. PMLR (2019)
13. Liu, F., Huang, X., Chen, Y., Suykens, J.A.: Random features for kernel approximation: a survey on algorithms, theory, and beyond. IEEE Trans. Pattern Anal. Mach. Intell. **44**(10), 7128–7148 (2021)
14. Mildenhall, B., Srinivasan, P.P., Tancik, M., Barron, J.T., Ramamoorthi, R., Ng, R.: NeRF: representing scenes as neural radiance fields for view synthesis. Commun. ACM **65**(1), 99–106 (2021)
15. Mormont, R., Geurts, P., Marée, R.: Comparison of deep transfer learning strategies for digital pathology. In: Proceedings of the IEEE Conference on Computer Vision and Pattern Recognition Workshops, pp. 2262–2271 (2018)
16. Munkhoeva, M., Kapushev, Y., Burnaev, E., Oseledets, I.: Quadrature-based features for kernel approximation. In: Advances in Neural Information Processing Systems, vol. 31 (2018)
17. Niemeyer, M., Geiger, A.: GIRAFFE: representing scenes as compositional generative neural feature fields. In: Proceedings of the IEEE/CVF Conference on Computer Vision and Pattern Recognition, pp. 11453–11464 (2021)
18. Oord, A.v.d., Li, Y., Vinyals, O.: Representation learning with contrastive predictive coding. arXiv preprint arXiv:1807.03748 (2018)
19. Park, W., Kim, D., Lu, Y., Cho, M.: Relational knowledge distillation. In: Proceedings of the IEEE/CVF Conference on Computer Vision and Pattern Recognition, pp. 3967–3976 (2019)
20. Passalis, N., Tzelepi, M., Tefas, A.: Probabilistic knowledge transfer for lightweight deep representation learning. IEEE Trans. Neural Netw. Learn. Syst. **32**(5), 2030–2039 (2020)
21. Peng, H., Pappas, N., Yogatama, D., Schwartz, R., Smith, N., Kong, L.: Random feature attention. In: International Conference on Learning Representations (2021)
22. Rahimi, A., Recht, B.: Random features for large-scale kernel machines. In: Advances in Neural Information Processing Systems, vol. 20 (2007)
23. Rawat, A.S., Chen, J., Yu, F.X.X., Suresh, A.T., Kumar, S.: Sampled softmax with random Fourier features. In: Advances in Neural Information Processing Systems, vol. 32 (2019)
24. Romero, A., Ballas, N., Kahou, S.E., Chassang, A., Gatta, C., Bengio, Y.: FitNets: hints for thin deep nets. arXiv preprint arXiv:1412.6550 (2014)
25. Tan, M., Le, Q.V.: EfficientNet: rethinking model scaling for convolutional neural networks. In: Proceedings of the 36th International Conference on Machine Learning, pp. 6105–6114 (2019)
26. Tancik, M., et al.: Fourier features let networks learn high frequency functions in low dimensional domains. In: Advances in Neural Information Processing Systems, vol. 33, pp. 7537–7547 (2020)
27. Vuong, T.T.L., Vu, Q.D., Jahanifar, M., Graham, S., Kwak, J.T., Rajpoot, N.: IMPaSh: a novel domain-shift resistant representation for colorectal cancer tissue classification. In: Karlinsky, L., Michaeli, T., Nishino, K. (eds.) ECCV 2022, Part III. LNCS, vol. 13803, pp. 543–555. Springer, Cham (2023). https://doi.org/10.1007/978-3-031-25066-8_31

28. Vuong, T.T., Song, B., Kim, K., Cho, Y.M., Kwak, J.T.: Multi-scale binary pattern encoding network for cancer classification in pathology images. IEEE J. Biomed. Health Inform. **26**(3), 1152–1163 (2021)
29. Wang, L., Lu, B., He, M., Wang, Y., Wang, Z., Du, L.: Prostate cancer incidence and mortality: global status and temporal trends in 89 countries from 2000 to 2019. Front. Public Health **10**, 811044 (2022)
30. Yu, F.X.X., Suresh, A.T., Choromanski, K.M., Holtmann-Rice, D.N., Kumar, S.: Orthogonal random features. In: Advances in Neural Information Processing Systems, vol. 29 (2016)

Semi-supervised Domain Adaptation for Automatic Quality Control of FLAIR MRIs in a Clinical Data Warehouse

Sophie Loizillon[1]([✉]), Olivier Colliot[1], Lydia Chougar[1], Sebastian Stroer[3], Yannick Jacob[2], Aurélien Maire[2], Didier Dormont[1,3], and Ninon Burgos[1]

[1] Sorbonne Université, Institut du Cerveau - Paris Brain Institute - ICM, CNRS, Inria, Inserm, AP-HP, Hôpital de la Pitié Salpêtrière, 75013 Paris, France
sophie.loizillon@gmail.com
[2] AP-HP, Innovation & Données – Département des Services Numériques, Paris, France
[3] AP-HP, Hôpital Pitié-Salpêtrière, DMU DIAMENT, Department of Neuroradiology, Paris, France

Abstract. Domain adaptation is a very useful approach to exploit the potential of clinical data warehouses, which gather a vast amount of medical imaging encompassing various modalities, sequences, manufacturers and machines. In this study, we propose a semi-supervised domain adaptation (SSDA) framework for automatically detecting poor quality FLAIR MRIs within a clinical data warehouse. Leveraging a limited number of labelled FLAIR and a large number of labelled T1-weighted MRIs, we introduce a novel architecture based on the well known Domain Adversarial Neural Network (DANN) that incorporates a specific classifier for the target domain. Our method effectively addresses the covariate shift and class distribution shift between T1-weighted and FLAIR MRIs, surpassing existing SSDA approaches by more than 10% points.

Keywords: Domain Adaptation · Deep Learning · Magnetic Resonance Imaging · Clinical Data Warehouse

1   Introduction

Machine learning algorithms aim to learn a model from training samples by minimising a cost function. However, for these models to work optimally, it is crucial that the training data and the test data share similar distributions. When the distribution of the training dataset differs from that of the test data, the performance of the model degrades. This would be the case when training an algorithm to rate the quality of T1-weighted (T1w) magnetic resonance (MR) images but

Supplementary Information The online version contains supplementary material available at https://doi.org/10.1007/978-3-031-45857-6_9.

applying it to another MRI sequence, e.g. fluid attenuated inversion recovery (FLAIR). To overcome this problem, various approaches have been proposed in the field of domain adaptation [1–4]. These approaches aim to develop robust models that can effectively adapt to various data distributions that differ from those on which they have been trained.

Domain adaptation has first been tackled in the literature in an unsupervised manner [1,5], with the objective of reducing the discrepancy between the source (e.g., T1w) and target (e.g., FLAIR) domains relying on labelled data from the source domain only. This is commonly referred to as unsupervised domain adaptation (UDA) Adversarial learning has achieved promising results by extracting domain invariant features. Adversarial learning attempts to extract relevant features to fool the domain classifier which aims to differentiate the source and target domains [5]. An effective adversarial learning approach is that of the Domain Adversarial Neural Network (DANN), which is composed of a shared feature extractor and a domain discriminator [1]. The domain discriminator aims to distinguish the source and target samples. At the same time, the feature extractor is trained to fool the domain discriminator in order to learn domain invariant features. While this method has proved its worth in recent years, Zhao et al. [6] warned that learning domain-independent features does not guarantee good generalisation to the target domain, particularly when the class distribution between domains differs.

In some cases, a few labelled samples are available from the target domain, which can help in improving the performance of the model. Saito et al. [2] introduced the notion of semi-supervised domain adaptation (SSDA), a variant of UDA where a limited number of labelled target samples is available. Specifically, they proposed a method called minimax entropy (MME). With the use of the mini-max paradigm, the presence of a small number of labelled target samples can considerably improve the performance of CNN models. The method consists of first estimating domain-invariant prototypes by maximising the entropy, and then clustering the features around the estimated prototypes, this time by minimising the same entropy. Several approaches focus on reducing the intra-domain divergence in the target domain [3,4]. Jiang et al. [3] designed an effective Bidirectional Adversarial Training (BiAT) to attract unaligned sub-distributions from the target domain to the corresponding source sub-distributions. While many existing SSDA approaches rely on adversarial techniques, there has been a recent interest in the use of contrastive learning. These methods, that show impressive success in self-supervised learning, aim to acquire an embedding space where similar pairs of samples are grouped together, while dissimilar pairs are pushed apart. In the context of domain adaptation, a common approach is to consider as similar pairs unlabelled target images and their corresponding augmented versions. On the other hand, dissimilar pairs are formed by combining unlabelled target images with labelled source images [7,8].

Due to the large variety of modalities and sequences, the field of medical imaging has turned its attention to the question of UDA and SSDA. Adversarial learning has been applied to various medical image classification and segmentation tasks [9–11]. An extension of the DANN to an encoder-decoder architecture is proposed in [9] and used for segmentation purposes in volume electron microscopy imaging. The network was initially trained using UDA techniques and subsequently fine-tuned on the target domain. A comparative study of the original DANN, a semi-supervised DANN, and fine-tuning techniques for white matter lesion segmentation is presented in [10]. Best results were obtained with the semi-supervised DANN. However, it was trained by passing all source and target samples through the same label classifier, which does not seem ideal especially when there is a distribution shift between source and target. Contrastive learning has also been used for domain adaptation in medical image analysis, for instance to reduce the gap between adult (source domain) and children (target domain) chest X-rays for automatic pneumonia diagnosis [11].

A sub-field of medical imaging where domain adaptation would be particularly necessary is that focusing on clinical routine data, such as those gathered in clinical data warehouses (CDWs). Having access to extensive and diverse clinical imaging data from multiple centres present an exceptional opportunity for the development of computational tools. To efficiently exploit CDWs, a model initially developed for a specific modality or sequence should have the ability to generalise to other modalities or sequences without the need to manually re-annotate large amounts of data. For example, recent studies have emphasised the importance of conducting automatic quality control (QC) on imaging data before deploying machine learning algorithms on CDWs [12]. To address this need, a framework was previously developed for the automatic QC of T1w brain MRIs. However, this framework, specific to the T1w sequence, required significant manual annotation efforts involving 5500 image annotations [13]. Our aim is to expand this framework to encompass FLAIR MRIs, while minimising the need for extensive new manual annotations using domain adaptation.

In this paper, we introduce a novel SSDA method for the automatic detection of poor quality FLAIR MRIs in a CDW while using a limited amount of labelled target data. In particular, we propose to incorporate a specific classifier for the target data into the well-known DANN [1]. We conduct a comparative analysis between our proposed method and the state-of-the-art SSDA architectures [2, 10,14].

2   Methods

2.1   Dataset Description

This work is built upon a large clinical routine dataset containing 3D T1w and FLAIR brain MR images of adult patients scanned in hospitals of the Greater Paris area (Assistance Publique-Hôpitaux de Paris [AP-HP]). The dataset is composed of 5500 T1w and 5858 FLAIR MRIs acquired on respectively 30 and 22 different scanners from four manufacturers (Siemens, GE, Philips, and Toshiba) which had been randomly selected from the CDW [13].

5500 T1w and 858 FLAIR MRIs were manually annotated with respect to the MRI quality. A score was given by two annotators to evaluate the noise, contrast and motion within the image on a three-point scale (0: no artefact, 1: moderate artefact, 2: severe artefact) [13]. In case of disagreement, the consensus score was chosen as the most severe of the two for T1w MRIs [13], while for the FLAIR images, both annotators reviewed the problematic images and agreed on a consensus, given that the number of manually annotated images was limited.

Based on the scores assigned to the motion, contrast and noise characteristics, we defined different tiers: *tier 1* represents good quality images (score of 0 for all characteristics), *tier 2* stands for medium quality images (score 1 for at least one characteristic and no score 2), and *tier 3* for bad quality images (score 2 for at least one characteristic). If the images did not contain full 3D T1w or FLAIR MRI of the whole brain, such as segmented tissue images or truncated images, they were classified as straight reject (*SR*). Figure 1 shows the distribution of the different tiers according to the modality in the form of a Sankey plot.

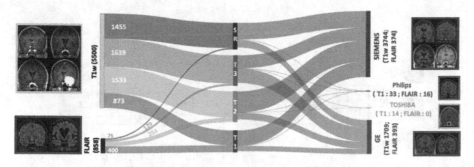

Fig. 1. Sankey plot analysis of labelled T1w and FLAIR MRIs highlighting the tier distribution for both sequences and the distribution of manufacturers across tiers.

2.2 Image Pre-processing

The T1w and FLAIR MRIs were pre-processed using Clinica and its {t1|flair} -linear pipeline [15] that relies on ANTs [16]. First, bias field correction was applied using the N4ITK method. Next, an affine registration to the MNI space was performed using a specific template for each of the T1w and FLAIR sequences. The images were then cropped to remove the background, resulting in $169 \times 208 \times 179$ images and 1 mm isotropic voxels.

2.3 Proposed Approach

We developed a semi-supervised approach based on the unsupervised DANN method to detect poor quality (i.e. *tier 3*) FLAIR in a clinical data warehouse.

The DANN [1], which aims to learn domain invariant features, was originally designed for unsupervised learning. The network is composed of a shared feature extractor (F) and two classifiers: a domain classifier (C_{dom}) and a label

predictor classifier. The goal of this architecture is to minimise the *label prediction loss* for accurate label prediction on the source domain and maximise the
domain confusion loss to align the feature distributions of the source and target
domains. Zhao et al. [6] showed that learning domain invariant features does
not necessarily guarantee the generalisation of the model to the target domain,
in particular when the class distributions change between the source and target domains. We adapted the DANN architecture to the context of SSDA by
adding a target label predictor classifier C_T, as shown in Fig. 2. Compared with
the approach of Sundaresan et al. [10], where all source and target data pass
through the same label classifier, the addition of a specific classifier to the target
data aims at handling the class distribution shift between the *tier 1/2* and *tier
3* classes for the T1w and FLAIR MRIs. While for the T1w sequence, the two
classes are slightly unbalanced (*tier 1/2*: 59% and *tier 3*: 41%), the imbalance
is much larger for the FLAIR (*tier 1/2*: 84% and *tier 3*: 16%).

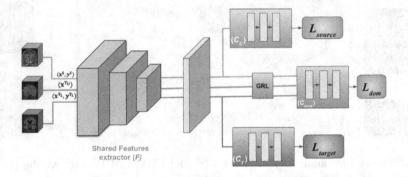

Fig. 2. Semi Supervised adapted DANN architecture composed of a domain invariant
feature extractor (F), a source label classifier (C_S), a target label classifier (C_T) and a
domain classifier (C_{dom}). A gradient reverse layer (GRL) multiplies the gradient by a
negative value when backpropagating to maximise the loss of the domain discriminator.

We denote the source dataset, consisting of labelled T1w MRIs, as $Ds = (x_i^S, y_i^S)_{i=1}^{N_S}$. For the target domain, we have two datasets: the labelled target samples $D_{T_L} = (x_i^{T_L}, y_i^{T_L})_{i=1}^{N_{T_L}}$ and the unlabelled target samples $D_{T_U} = (x_i^{T_U})_{i=1}^{N_{T_U}}$.
We will refer to the labelled data from the two domains using the following
notation $D_L = D_S \cup D_{T_L}$. The overall loss for the semi supervised DANN is

$$L = \underbrace{L_{pred}(F, C_S, C_T, D_L)}_{\text{Label Prediction Loss}} - \lambda \cdot \underbrace{L_{dom}(F, C_{dom}, D_L, D_{T_U}, d)}_{\text{Domain Confusion Loss}} \quad (1a)$$

where

$$L_{\text{pred}} = -\underbrace{\sum_{k=1}^{K}(y_i^S)_k \log(C_S(F(x_i^S))_k}_{\text{Source Label Prediction}} - \underbrace{\sum_{k=1}^{K}(y_i^{T_L})_k \log(C_T(F(x_i^{T_L}))_k}_{\text{Target Label Prediction}} \qquad (1b)$$

$$L_{\text{dom}} = d_i \log \frac{1}{C_{dom}(F(x_i))} + (1 - d_i) \log \frac{1}{1 - C_{dom}(F(x_i))} \qquad (1c)$$

with d_i the domain label of the i-th sample, which indicates whether it is a T1w (source domain) or a FLAIR (target domain), and $K = 2$ is the number of classes for the source and target label classifier (i.e., task *tier1/2* vs *tier3*).

As in the classical DANN [1], the model is initialised using a pre-trained model without the domain classifier. This model is usually trained on the source domain that has been labelled. Here we pre-trained the network using all the labelled data from the source and target domains. The training procedure is detailed in Algorithm 1. We structured each iteration to include three mini-batches. A first mini-batch contained labelled samples from the source domain (b_S), another consisted of labelled samples from the target domain (b_{T_L}), and the final mini-batch comprised unlabelled samples from the target domain (b_{T_U}). This arrangement ensured that a target labelled data sample was present in every batch, effectively influencing the training process.

Algorithm 1. Proposed semi-supervised DANN

Require:
1: Source samples $D_S = \{(x_i^S, y_i^S)\}_{i=1}^{N_S}$
2: Target labelled samples $D_{T_L} = \{(x_i^{T_L}, y_i^{T_L})\}_{i=1}^{N_{T_L}}$
3: Target unlabelled samples $D_{T_U} = \{(x_i^{T_U})\}_{i=1}^{N_{T_U}}$
4: Hyperparameters : λ, learning rate η
5: Pre-trained model with all labelled samples (D_S, D_{T_L}) and $\lambda = 0$
6: **for** each epoch **do**
7:     **for** each b_S, b_{T_L}, b_{T_U} **do**
8:        # **Forward pass**
9:        $\hat{y}^S \leftarrow C_S(F(b_S, \theta_F), \theta_{C_S})$
10:        $\hat{y}^T \leftarrow C_T(F(b_{T_L}, \theta_F), \theta_{C_T})$
11:        $\hat{d} \leftarrow C_{\text{dom}}(F(b_S + b_{T_L} + b_{T_U}, \theta_F), \theta_{C_{\text{dom}}})$
12:        # **Loss computation**
13:        $L \leftarrow L_S(\hat{y}^S, y^S) + L_{T_L}(\hat{y}^{T_L}, y^{T_L}) + \lambda L_{\text{dom}}(\hat{d}, d)$
14:        # **Backward pass & model weight update**
15:        $\theta_F \leftarrow \theta_F - \eta \times \nabla_{\theta_F} L$
16:        $\theta_{C_S} \leftarrow \theta_{C_S} - \eta \times \nabla_{\theta_{C_S}} L$
17:        $\theta_{C_T} \leftarrow \theta_{C_T} - \eta \times \nabla_{\theta_{C_T}} L$
18:     **end for**
19: **end for**

2.4 Experiments

We aimed to assess the ability of the proposed and different existing SSDA approaches to detect bad quality FLAIR MRIs, which corresponds to the classification task *tier 1/2* vs *tier 3*. Before starting the experiments, we excluded all the *SR* images that contained MRIs which were not full 3D MRIs of the whole brain (1455 T1w and 75 FLAIR), e.g., truncated images or segmented brain tissue images. We built the FLAIR MRIs test set by randomly selecting 480 manually annotated images while ensuring they shared the same distribution of tiers, manufacturers, and field strengths as the images in the training/validation set. The remaining 303 labelled FLAIR MRIs were split into training and validation using a 5-fold cross validation. In a similar manner, the 5000 unlabelled FLAIR MRIs were divided into training and validation sets, taking into account the same distribution of manufacturers and field strengths. As for the T1w dataset composed of 3660 labelled samples, we employed the identical split as Bottani et al. [13], where the images were split between training and validation using a 5-fold cross validation respecting the same tiers distribution. We conducted a comparative study between our proposed approach and three well-known SSDA methods that use labelled data from the source and target domains to jointly train a network, and unlabelled samples to regularise it: a semi-supervised DANN without target label classifier [10], mini-max entropy (MME) [2] and entropy minimisation (ENT) [14].

MME [2] extracts discriminating and domain-invariant features using unlabelled target data by estimating domain-invariant prototypes. Domain adaptation is performed by first maximising the entropy of unlabelled target data according to the classifier to estimate prototypes that are invariant to the domain. Then, the entropy is minimised regarding to the feature extractor to make target features well-clustered around the prototypes.

ENT [14] relies on entropy minimisation using labelled source and target data, along with unlabelled target data. The entropy is computed on the unlabelled target samples, and the network is trained to minimise this entropy. In contrast to MME, ENT does not involve a maximisation process.

All the experiments were conducted using the ClinicaDL software, implemented in PyTorch [17]. The Conv5FC3 network, consisting of five convolutional neural network layers and three fully connected layers, was used for every experiment as its performance was as good as that of more sophisticated CNN architectures (Inception, ResNet) on the T1w MRI QC task [13].

3 Results

The main results for the detection of bad quality MRIs (*tier 3*) within the CDW are shown in Table 1. For each method, we report the results on the independent T1w and FLAIR test sets. We first trained the baseline for the source domain (Baseline T1w) using the 3660 labelled T1w MRIs. The model exhibited excellent balanced accuracy (BA) on the T1w test set (83.51%) but the performance on the FLAIR test set dramatically dropped (50.06%). We performed the same experiments training on the target domain using the 303 labelled FLAIR MRIs (Baseline FLAIR). We obtained a low BA of 68.23% on the FLAIR test set. Finally, the baseline trained with all the labelled data (Baseline T1 + FLAIR) achieved an excellent BA on the T1w but a low BA on FLAIR. With regard to the results obtained for the different SSDA methods, the approach we have proposed outperforms the DANN, MME and ENT by 10, 13 and 14 percent points respectively on FLAIR MRIs, reaching a BA of 76.81%. The evaluation of the different models, including specificity and sensitivity metrics, is presented in Table 1 of the supplementary material.

Table 1. Results for the detection of bad quality images (*tier 3*) within the T1w and FLAIR test sets from the CDW. We report the mean and empirical standard deviation across the five folds of the balanced accuracy (BA), which is defined as the mean of the specificity and sensitivity. For Manual Annotation, the BA corresponds to the average BA of the two annotators with the consensus.

Approaches	N of S_l	N of T_l	N of T_u	T1w BA	FLAIR BA
Manual Annotation	3660	303	–	91.56	86.54
Baseline T1w	3660	0	0	83.51 ± 0.93	50.06 ± 1.71
cre Baseline FLAIR	0	303	0	48.20 ± 0.81	69.90 ± 2.38
Baseline T1w + FLAIR	3660	303	0	82.37 ± 0.85	58.05 ± 4.45
DANN [10]	3660	303	5000	81.97 ± 1.39	66.91 ± 2.37
ENT [14]	3660	303	5000	80.07 ± 1.99	62.21 ± 4.45
MME [2]	3660	303	5000	77.39 ± 3.82	63.31 ± 4.16
Proposed Approach	3660	303	5000	80.59 ± 1.62	76.81 ± 0.68

The observed results of the T1w baseline are in line with expectations, considering the training was conducted only on T1w images. These T1w images possess different intensity distributions when compared to the target distribution of FLAIR MRIs ($P_S(X, Y) \neq P_T(X, Y)$), which consequently led to a decline in performance when applied to FLAIR samples. Similarly, the FLAIR baseline demonstrated poor results on T1w MRIs due to the covariate shift between the two sequences. Furthermore, the limited amount of manually annotated FLAIR images (303 samples) contributed to a low balanced accuracy (BA: 68%) on the FLAIR test set. Finally, the T1w+FLAIR baseline exhibited satisfying results on the T1w test set. However, since the training labels are highly dominated by T1w images, it led to a low BA on the FLAIR test set. This underlines the significance of employing SSDA techniques to develop models that are robust to different MRI sequences.

The poor results obtained from the DANN, MME and ENT on the target domain (BA < 67%) underline the fact that learning domain invariant features is not sufficient in particular in the case of class distribution shift [6]. Indeed, while the distribution between labels was slightly unbalanced (2406 vs 1639 images) for the source domain, it was extremely unbalanced for the target domain (654 vs 129). This difference in class distributions between the T1w (source) and FLAIR (target) MRIs explains the poor performance of these methods. With our proposed approach of incorporating a second label classifier dedicated to the target modality, we were able to overcome these limitations and achieved an important improvement of more than 10 percent points. Thus, this model will be applicable to filter and select good quality FLAIR MRIs from the AP-HP CDW, enabling users to take advantage of these sequences for conducting further studies.

4   Conclusion

In this paper, we propose a new SSDA architecture based on the original DANN [1] for the detection of bad quality FLAIR MRIs in a CDW. By adding a second label classifier specifically for the target domain, we were able to overcome the covariate shift and the class distributions shift between the source and target domains. We achieved a balanced accuracy of 76.8% on the FLAIR test set and outperforms the DANN, MME and ENT by 10, 13 and 14 percent points respectively.

References

1. Ganin, Y., et al.: Domain-adversarial training of neural networks. J. Mach. Learn. Res. **17**(1), 2096–2130 (2016)
2. Saito, K., Kim, D., Sclaroff, S., Darrell, T., Saenko, K.: Semi-supervised domain adaptation via minimax entropy. In: Proceedings of the IEEE/CVF International Conference on Computer Vision, pp. 8050–8058 (2019)
3. Jiang, P., Wu, A., Han, Y., Shao, Y., Qi, M., Li, B.: Bidirectional adversarial training for semi-supervised domain adaptation. IJCA **I**, 934–940 (2020)
4. Kim, T., Kim, C.: Attract, perturb, and explore: learning a feature alignment network for semi-supervised domain adaptation. In: Vedaldi, A., Bischof, H., Brox, T., Frahm, J.-M. (eds.) ECCV 2020. LNCS, vol. 12359, pp. 591–607. Springer, Cham (2020). https://doi.org/10.1007/978-3-030-58568-6_35
5. HassanPour Zonoozi, M., Seydi, V.: A survey on adversarial domain adaptation. Neural Process. Lett. **55**, 1–41 (2022)
6. Zhao, H., Des Combes, R.T., Zhang, K., Gordon, G.: On learning invariant representations for domain adaptation. In: International Conference on Machine Learning, pp. 7523–7532. PMLR (2019)
7. Singh, A.: CLDA: contrastive learning for semi-supervised domain adaptation. In: Advances in Neural Information Processing Systems, vol. 34, pp. 5089–5101 (2021)
8. Thota, M., Leontidis, G.: Contrastive domain adaptation. In: Proceedings of the IEEE/CVF Conference on Computer Vision and Pattern Recognition, pp. 2209–2218 (2021)
9. Roels, J., Hennies, J., Saeys, Y., Philips, W., Kreshuk, A.: Domain adaptive segmentation in volume electron microscopy imaging. In: 2019 IEEE 16th International Symposium on Biomedical Imaging (ISBI 2019), pp. 1519–1522. IEEE (2019)

10. Sundaresan, V., Zamboni, G., Dinsdale, N.K., Rothwell, P.M., Griffanti, L., Jenkinson, M.: Comparison of domain adaptation techniques for white matter hyperintensity segmentation in brain MR images. Med. Image Anal. **74**, 102215 (2021)
11. Feng, Y., et al.: Contrastive domain adaptation with consistency match for automated pneumonia diagnosis. Med. Image Anal. **83**, 102664 (2023)
12. Bottani, S., et al.: Evaluation of MRI-based machine learning approaches for computer-aided diagnosis of dementia in a clinical data warehouse. Preprint (2023). https://hal.science/hal-03656136
13. Bottani, S., et al.: Automatic quality control of brain T1-weighted magnetic resonance images for a clinical data warehouse. Med. Image Anal. **75**, 102219 (2022)
14. Grandvalet, Y., Bengio, Y.: Semi-supervised learning by entropy minimization. In: Saul, L., Weiss, Y., Bottou, L. (eds.) Advances in Neural Information Processing Systems, vol. 17. MIT Press (2004)
15. Routier, A., et al.: Clinica: an open-source software platform for reproducible clinical neuroscience studies. Front. Neuroinform. **15**, 689675 (2021)
16. Avants, B.B., Tustison, N.J., Stauffer, M., Song, G., Wu, B., Gee, J.C.: The insight ToolKit image registration framework. Frontiers in Neuroinformatics **8**, 44 (2014)
17. Thibeau-Sutre, E., et al.: ClinicaDL: an open-source deep learning software for reproducible neuroimaging processing. Comput. Methods Programs Biomed. **220**, 106818 (2022)

Towards Foundation Models Learned from Anatomy in Medical Imaging via Self-supervision

Mohammad Reza Hosseinzadeh Taher[1], Michael B. Gotway[2], and Jianming Liang[1]([✉])

[1] Arizona State University, Tempe, AZ 85281, USA
{mhossei2,jianming.liang}@asu.edu
[2] Mayo Clinic, Scottsdale, AZ 85259, USA
Gotway.Michael@mayo.edu

Abstract. Human anatomy is the foundation of medical imaging and boasts one striking characteristic: its hierarchy in nature, exhibiting two intrinsic properties: (1) *locality*: each anatomical structure is morphologically distinct from the others; and (2) *compositionality*: each anatomical structure is an integrated part of a larger whole. We envision a foundation model for medical imaging that is *consciously* and *purposefully* developed upon this foundation to gain the capability of "understanding" human anatomy and to possess the fundamental properties of medical imaging. As our first step in realizing this vision towards foundation models in medical imaging, we devise a novel self-supervised learning (SSL) strategy that exploits the hierarchical nature of human anatomy. Our extensive experiments demonstrate that the SSL pretrained model, derived from our training strategy, not only outperforms state-of-the-art (SOTA) fully/self-supervised baselines but also enhances annotation efficiency, offering potential few-shot segmentation capabilities with performance improvements ranging from 9% to 30% for segmentation tasks compared to SSL baselines. This performance is attributed to the significance of *anatomy comprehension* via our learning strategy, which encapsulates the intrinsic attributes of anatomical structures—*locality* and *compositionality*—within the embedding space, yet overlooked in existing SSL methods. All code and pretrained models are available at GitHub.com/JLiangLab/Eden.

Keywords: Self-supervised Learning · Learning from Anatomy

1 Introduction and Related Works

Foundation models [4], such as GPT-4 [22] and DALL.E [23], pretrained via self-supervised learning (SSL), have revolutionized natural language processing (NLP) and radically transformed vision-language modeling, garnering significant public media attention [18]. But, despite the development of numerous SSL

Supplementary Information The online version contains supplementary material available at https://doi.org/10.1007/978-3-031-45857-6_10.

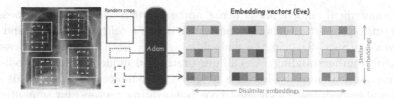

Fig. 1. Existing SSL methods lack capabilities of "understanding" the foundation of medical imaging—human anatomy. We believe that a foundation model must be able to transform each pixel in an image (e.g., a chest X-ray) into semantics-rich numerical vectors, called embeddings, where different anatomical structures (indicated by different colored boxes) are associated with different embeddings, and the same anatomical structures have (nearly) identical embeddings at all resolutions and scales (indicated by different box shapes) across patients. Inspired by the hierarchical nature of human anatomy (Fig. 6 in Appendix), we introduce a novel SSL strategy to learn anatomy from medical images (Fig. 2), resulting in embeddings (Eve), generated by our pretrained model (Adam), with such desired properties (Fig. 4 and Fig. 8 in Appendix).

methods in medical imaging, their success in this domain lags behind their NLP counterparts. What causes these striking differences? We believe that this is because the SSL methods developed for NLP have proven to be powerful in capturing the underlying structures (foundation) of the English language; thus, a number of intrinsic properties of the language emerge naturally, as demonstrated in [19], while the existing SSL methods lack such capabilities to appreciate the foundation of medical imaging—human anatomy. Therefore, this paper is seeking to answer a fundamental question: *How to learn foundation models from human anatomy in medical imaging?*

Human anatomy exhibits natural hierarchies. For example, the lungs are divided into the right and left lung (see Fig. 6 in Appendix) and each lung is further divided into lobes, two on the left and three on the right lung. Each lobe is further subdivided into segments, each containing pulmonary arteries, veins, and bronchi which branch in predictable, dichotomous fashion. Consequently, anatomical structures have two important properties: *locality*: each anatomical structure is morphologically distinct from others; *compositionality*: each anatomical structure is an integrated part of a larger whole. Naturally, a subquestion is *how to exploit the anatomical hierarchies for training foundation models?* To this end, we devise a novel SSL training strategy, which is *hierarchical, autodidactic,* and *coarse*, resulting in a pretrained model, which is *versatile,* and leading to anatomical embedding, which is *dense* and *semantics-meaningful.* Our training strategy is *hierarchical* because it decomposes and perceives the anatomy progressively in a coarse-to-fine manner (Sect. 2.1); *autodidactic* because it learns from anatomy through self-supervision, thereby requiring no anatomy labeling (Sect. 2); and *coarse* because it generates dense anatomical embeddings without relying on pixel-level training (Sect. 3, ablation 1). The pretrained model is *versatile* because it is strong in generality and adaptability, resulting in performance boosts (Sect. 3.1) and annotation efficiency (Sect. 3.2) in myriad tasks. The generated anatomical embedding is *dense* and *semantics-rich* because it possesses two intrinsic properties of anatomical structures, *locality* (Sect. 3.3) and *compositionality* (Sect. 3.4), in the embedding space, both of which are essential for anatomy understanding. We call our pretrained model **Adam** (*a*utodidactic *d*ense *a*natomical *m*odels) because it learns autodidactically and yields dense anatomical embedding, nicknamed **Eve** (*e*mbedding *v*ectors) for semantic richness (Fig. 1). We further coin our project site **Eden** (*e*nvironment for *d*ense *e*mbeddings and *n*etworks), where all code, pretrained Adam and Eve are placed.

In summary, we make the following contributions: (**1**) A novel self-supervised learning strategy that progressively learns anatomy in a coarse-to-fine manner via hierarchical contrastive learning; (**2**) A new evaluation approach that facilitates analyzing the interpretability of deep models in anatomy understanding by measuring the locality and compositionality of anatomical structures in embedding space; and (**3**) A comprehensive and insightful set of experiments that evaluate Adam for a wide range of 9 target tasks, involving fine-tuning, few-shot learning, and investigating semantic richness of Eve in anatomy understanding.

Related Works: (i) Self-supervised learning methods, particularly contrastive techniques [2,16], have shown great promise in medical imaging [12,25]. But, due to their focus on image-level features, they are sub-optimal for dense recognition tasks [28]. Recent works [10,13] empower contrastive learning with more discriminative features via using the diversity in the local context of medical images. In contrast to them, which overlook anatomy hierarchies in their learning objectives, Adam exploits the hierarchical nature of anatomy to learn semantics-rich dense features. **(ii) Anatomy learning** methods integrate anatomical cues into their SSL objectives. But, GLC [6] requires spatial correspondence across images, limiting its scalability to non-aligned images. Although TransVW [11], SAM [31], and Alice [15] relax this requirement, they neglect hierarchical anatomy relations, offering no compositionality. By contrast, Adam learns consistent anatomy features without relying on spatial alignment across images (see Fig. 7 in Appendix) and captures both local and global contexts hierarchically to offer both locality and compositionality. **(iii) Hierarchical SSL** methods exploit transformers' self-attention to model dependencies among image patches. But, they fail to capture anatomy relations due to inefficient SSL signals that contrast similar anatomical structures [26] or disregard relations among images [29,30]. Adam goes beyond architecture design by introducing a learning strategy that decomposes anatomy into a hierarchy of parts for coarse-to-fine anatomy learning, and avoids semantic collision in its supervision signal.

2   Method

Our self-supervised learning strategy, depicted in Fig. 2, aims to exploit the hierarchical nature of human anatomy in order to capture not only generic but also semantically meaningful representations. The main intuition behind our learning strategy is the principle of totality in *Gestalt* psychology: humans commonly first recognize the prominent objects in an image (e.g., lungs) and then *gradually*

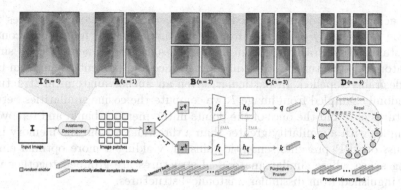

Fig. 2. Our SSL strategy gradually decomposes and perceives the anatomy in a coarse-to-fine manner. Our Anatomy Decomposer (AD) decomposes the anatomy into a hierarchy of parts with granularity level $n \in \{0, 1, ..\}$ at each training stage. Thus, anatomical structures of finer-grained granularity will be incrementally presented to the model as the input. Given image I, we pass it to AD to get a random anchor x. We augment x to generate two views (positive samples), and pass them to two encoders to get their features. To avoid semantic collision in training objective, our Purposive Pruner removes *semantically* similar anatomical structures across images to anchor x from the memory bank. Contrastive loss is then calculated using positive samples' features and the pruned memory bank. The figure shows pretraining at $n = 4$.

recognize smaller details based on prior knowledge about that object (e.g., each lung is divided into lobes) [24]. Inspired by this principle, we propose a training strategy, which decomposes and perceives the anatomy progressively in a coarse-to-fine manner, aiming to learn both anatomical (local and global) contextual information and also the relative hierarchical relationships among anatomical structures. Our framework is comprised of two key components:

(1) Anatomy Decomposer (AD) is responsible for decomposing relevant anatomy into a hierarchy of anatomical structures to guide the model to learn hierarchical anatomical relationships in images. The AD component takes two inputs: an image I and an anatomy granularity level n, and generates a random anatomical structure instance x. We generate anatomical structures at desired granularity level n in a recursive manner. Given an image I, we first split it vertically into two halves (A in Fig. 2). Then, we iteratively alternate between horizontally and vertically splitting the resulting image parts until we reach the desired granularity level (B, C, D in Fig. 2). This process results in 2^n image patches $\{x_i\}_{i=1}^{2^n}$. In this set, we randomly sample an instance x, which will be used as the input for training the model. As such, during the pretraining, anatomical structures at various granular levels are generated and present to the model.

(2) Purposive Pruner (PP) is responsible for compeling the model to comprehend anatomy more effectively via learning a wider range of distinct anatomical structures. Intuitively, similar anatomical structures (e.g. ribs or disks) should have similar embeddings, while also their finer-grained constituent parts (e.g. different ribs or disks) have (slightly) different embeddings.

To achieve such desired embedding space, the anatomical structures need to be intelligently contrasted from each other. Our PP module, in contrast to standard contrastive learning approaches, identifies *semantically* similar anatomical structures in the embedding space and prevents them from being undesirably repelled. In particular, given an anchor anatomical structure x randomly sampled from image I, we compute the cosine similarities between features of x and the ones of the points in the memory bank, and remove the samples with a similarity greater than a threshold γ from the memory bank. Thus, our PP prevents semantic collision, yielding a more optimal embedding space where similar anatomical structures are grouped together while distinguished from dissimilar anatomical structures.

Overall Training. Our framework consists of two twin backbones f_θ and f_ξ, and projection heads h_θ and h_ξ. f_θ and h_θ are updated by back-propagation, while f_ξ and h_ξ are updated by exponential moving average (EMA) of f_θ and h_θ parameters, respectively. We use a memory bank to store the embeddings of negative samples MB $= \{k_i\}_{i=1}^{K}$, where K is the memory bank size. For learning anatomy in a coarse-to-fine manner, we progressively increase the anatomical structures granularity. Thus, at each training stage, anatomical structures with granularity level $n \in \{0, 1, ..\}$ will be presented to the model. Given input image I and data granularity level n, we pass them to our AD to get a random anatomical structure x. We apply an augmentation function $T(.)$ on x to generate two views x_q and x_k, which are then processed by backbones and projection heads to generate latent features $q = h_\theta(f_\theta(x_q))$ and $k = h_\xi(f_\xi(x_k))$. Then, we pass q and MB to our PP to remove false negative samples for anchor x, resulting in pruned memory bank MB$_{\text{pruned}}$, which is used to compute the InfoNCE [7] loss

$$\mathcal{L} = -log\frac{exp(q \cdot k/\tau)}{exp(q \cdot k/\tau) + \sum_{i=1}^{K'} exp(q \cdot k_i/\tau)}, \text{ where } \tau \text{ is a temperature hyperparameter, } K'$$

is size of MB$_{\text{pruned}}$, and $k_i \in$ MB$_{\text{pruned}}$. Our AD module enables the model to first learn anatomy at a coarser-grained level, and then use this acquired knowledge as effective contextual clues for learning more fine-grained anatomical structures, reflecting anatomical structures compositionality in its embedding space. Our PP module enables the model to learn a semantically-structured embedding space that preserves anatomical structures locality by removing semantic collision from the model's learning objective. The pretrained model derived by our training strategy (**Adam**) can not only be used as a basis for myriad target tasks via adaptation (*fine-tuning*), but also its embedding vectors (**Eve**) show promises to be used standalone *without* adaptation for other tasks like landmark detection.

3 Experiments and Results

Pretraining and Fine-Tuning Settings: We use unlabeled training images of ChestX-ray14 [27] and EyePACS [8] for pretraining and follow [7] in pretraining settings: SGD optimizer with an initial learning rate of 0.03, weight decay

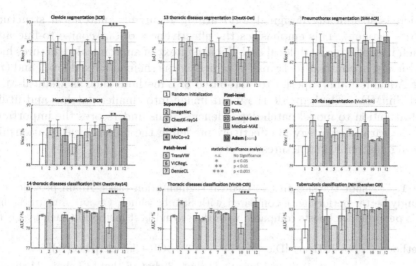

Fig. 3. Adam provides superior performance over fully/self-supervised methods. All SSL methods are pretrained on ChestX-ray14 dataset. Statistical significance analysis ($p < 0.05$) was conducted between Adam and the top SSL baseline in each task.

1e-4, SGD momentum 0.9, cosine decaying scheduler, and batch size 256. The input anatomical structures are resized to 224×224; augmentations include random crop, color jitter, Gaussian blur, and rotation. We use data granularity level (n) up to 4 and pruning threshold $\gamma = 0.8$ (ablation in Appendix). Following [10,16], we adopt ResNet-50 as the backbone. For fine-tuning, we (1) use the pretrained encoder followed by a task-specific head for classification tasks, and a U-Net network for segmentation tasks where the encoder is initialized with the pretrained backbone; (2) fine-tune all downstream model's params; (3) run each method 10 times on each task and report statistical significance analysis.

Downstream Tasks and Baselines: We evaluate Adam on a myraid of 9 tasks on ChestX-ray14 [27], Shenzhen [14], VinDr-CXR [20], VinDR-Rib [21], SIIM-ACR [1], SCR [9], ChestX-Det [17], and DRIVE [5], covering various challenging tasks, diseases, and organs. We compare Adam with SOTA *image-* (MoCo-v2 [7]), *patch-* (TransVW [11], VICRegL [3], DenseCL [28]), and *pixel-level* (PCRL [32], DiRA [10], Medical-MAE [29], SimMIM [30]) SSL methods.

1) Adam provides generalizable representations for a variety of tasks.

To showcase the significance of anatomy learning via our SSL approach and its impact on representation learning, we compare transfer learning performance of Adam to 8 recent SOTA SSL methods with diverse objectives, as well as 2 fully-supervised models pretrained on ImageNet and ChestX-ray14 datasets, in 8 downstream tasks. As seen in Fig. 3, (*i*) our Adam consistently outperforms the SOTA dense SSL methods (VICRegL & DenseCL) as well as the SOTA medical SSL methods (PCRL & DiRA), and achieves superior or comparable performance compared to fully-supervised baselines; (*ii*) our Adam demonstrates a significant performance improvement over TransVW, which is specifically designed for learning recurring anatomical structures across patients. This emphasizes the effectiveness of our coarse-to-fine approach in capturing both local and global context of anatomical structures hierarchically, in contrast to TransVW which learns them at a fixed level; and (*iii*) our Adam remains superior to ViT-based SSL methods such as Medical-MAE and SimMIM, which divide the input image into smaller patches and utilize self-attention to model patch dependencies. This underscores the importance of our learning strategy in effectively modeling the hierarchical relationships among anatomical structures.

Table 1. Few-shot transfer on two medical segmentation tasks. Adam provides outstandingly better performance compared with SSL baselines. Green numbers show Adam's performance boosts compared with the second-best method in each task/shot.

Method	SCR-Heart [Dice(%)]				SCR-Clavicle [Dice(%)]			
	3-shot	6-shot	12-shot	24-shot	3-shot	6-shot	12-shot	24-shot
MoCo-v2	44.84	59.97	69.90	79.69	23.77	29.24	38.07	44.47
DenseCL	64.88	74.43	75.79	80.06	36.43	51.31	63.03	69.13
DiRA	63.76	64.47	76.10	81.42	31.42	38.59	66.81	73.06
Adam (ours)	**84.35**	**86.70**	**89.79**	**90.45**	**66.69**	**79.41**	**83.96**	**84.76**
	(↑19)	(↑12)	(↑14)	(↑9)	(↑30)	(↑28)	(↑17)	(↑12)

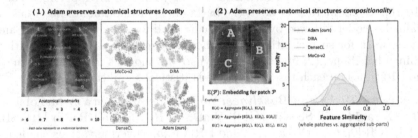

Fig. 4. Adam preserves locality and compositionality properties, which are intrinsic to anatomical structures and critical for understanding anatomy, in its embedding space.

2) Adam enhances annotation efficiency, revealing promise for few-shot learning. To dissect robustness of our representations, we compare Adam with top-performing SSL methods from each baseline group, based on Fig. 3, in limited data regimes. We conduct experiments on Heart and Clavicle segmentation tasks, and fine-tune the pretrained models using a few shots of labeled data (3, 6, 12, and 24) randomly sampled from each dataset. As seen in Table 1, Adam not only demonstrates superior performance against baselines by a large margin (Green *nums.*) but also maintains consistent behavior with minimal performance drop as labeled data decreases, compared to baselines. We attribute Adam's superior representations over baselines, as seen in Fig. 3 and Table 1, to its ability to learn the anatomy by preserving locality and compositionality of anatomical structures in its embedding space, as is exemplified in the following.

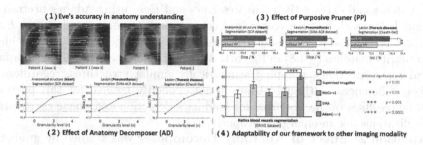

Fig. 5. Ablation studies on (1) Eve's accuracy in anatomy understanding, (2) effect of anatomy decomposer, (3) effect of purposive pruner, and (4) adaptability of our framework to other imaging modalities.

3) Adam preserves anatomical structures locality. We investigate Adam's ability to reflect *locality* of anatomical structures in its embedding space against existing SSL baselines. To do so, we (1) create a dataset of 1,000 images (from ChestX-ray14 dataset) with 10 distinct anatomical landmarks *manually* annotated by human experts in each image, (2) extract 224 × 224 patches around each landmark across images, (3) extract latent features of each landmark instance using each pretrained model under study and then pass them through a global average pooling layer, and (4) visualize the features by using t-SNE. As seen in Fig. 4.1, existing SSL methods lack the ability in discriminating different anatomical structures, causing ambiguous embedding spaces. In contrast, Adam excels in distinguishing various anatomical landmarks, yielding well-separated clusters in its embedding space. This highlights Adam's ability to learn a rich semantic embedding space where distinct anatomical structures have unique embeddings, and identical structures share near-identical embeddings across patients.

4) Adam preserves anatomical structures compositionality. The embedding of a whole should be equal or close to the sum of the embedding of its each part (see $\mathbb{E}(\mathcal{P})$ examples in Fig. 4.2). To investigate Adam's ability to reflect *compositionality* of anatomical structures in its embedding space against existing SSL baselines, we (1) extract random patches from test images of ChestX-ray14, and decompose each patch into 2, 3, or 4 non-overlapping sub-patches, (2) resize each extracted patch and its sub-patches to 224 × 224 and then extract their features using each pretrained model under study, (3) compute cosine similarity between the embedding of each patch and the aggregate of the embeddings of its sub-patches, and (4) visualize the similarity distributions with Gaussian kernel density estimation (KDE). As seen in Fig. 4.2, Adam's distribution is not only narrower and taller than baselines, but also the mean of similarity value between embedding of whole patches and their aggregated sub-parts is closer to 1.

Ablation 1: Eve's accuracy in anatomy understanding was studied by visualizing dense correspondence between (i) an image and its augmented views and (ii) different images. Given two images, we divide them into grids of patches and extract their features Eve_1 and Eve_2 using Adam's pretrained model. For each feature vector in Eve_1, we find its correspondence in Eve_2 based on highest *cosine* similarity; for clarity, we show some of the high-similarity matches (≥ 0.8) in Fig. 5.1. As seen, Eve has accurate dense anatomical representations, mapping semantically similar structures, regardless of their differences. Although Adam is not explicitly trained for this purpose, these results show its potential for landmark detection and image registration applications, as an emergent property.

Ablation 2: Effect of Anatomy Decomposer was studied by gradually increasing pretraining data granularity from coarse-grained anatomy ($n = 0$) to finer levels (up to $n = 4$) and fine-tuning the models on downstream tasks. As seen in Fig. 5.2, gradual increment of data granularity consistently improves the performance across all tasks. This suggests that our coarse-to-fine learning strategy deepens the model's anatomical knowledge.

Ablation 3: Effect of Purposive Pruner was studied by comparing a model with and *without* PP (i.e. contrasting an anchor with all negative pairs in the memory bank) during pretraining. Figure 5.3 shows PP leads to significant performance boosts across all tasks, highlighting its key role in enabling the model to capture more discriminative features by removing noisy contrastive pairs.

Ablation 4: Adaptability of our framework to other imaging modalities was explored by utilizing fundoscopy photography images in EyePACS as pretraining data, which possess complex structures due to the diverse variations in retinal anatomy. As depicted in Fig. 5.4, Adam provides superior performance by 1.4% ($p < 0.05$) in the blood vessel segmentation task compared to the top-performing SSL methods that also leverage the same pretraining images. This highlights the importance of effectively learning the anatomy and also showcases the potential applicability of our method to various imaging modalities.

4   Conclusion and Future Work

A key contribution of ours lies in crafting a novel SSL strategy that underpins the development of powerful self-supervised models foundational to medical imaging via learning anatomy. Our training strategy progressively learns anatomy in a coarse-to-fine manner via hierarchical contrastive learning. Our approach yields highly generalizable pretrained models and anatomical embeddings with essential properties of *locality* and *compositionality*, making them semantically meaningful for anatomy understanding. In future, we plan to apply our strategy to provide *dense anatomical models* for major imaging modalities and protocols.

References

1. SIIM-ACR pneumothorax segmentation (2019). https://www.kaggle.com/c/siim-acr-pneumothorax-segmentation/
2. Azizi, S., et al.: Big self-supervised models advance medical image classification. In: Proceedings of the IEEE/CVF International Conference on Computer Vision, pp. 3478–3488 (2021)
3. Bardes, A., Ponce, J., LeCun, Y.: VICRegl: self-supervised learning of local visual features. In: Advances in Neural Information Processing Systems, vol. 35, pp. 8799–8810 (2022)
4. Bommasani, R., et al.: On the opportunities and risks of foundation models. ArXiv (2021). https://crfm.stanford.edu/assets/report.pdf
5. Budai, A., Bock, R., Maier, A., Hornegger, J., Michelson, G.: Robust vessel segmentation in fundus images. Int. J. Biomed. Imaging **2013**, 154860 (2013)
6. Chaitanya, K., Erdil, E., Karani, N., Konukoglu, E.: Contrastive learning of global and local features for medical image segmentation with limited annotations. In: Advances in Neural Information Processing Systems, vol. 33, pp. 12546–12558 (2020)
7. Chen, X., Fan, H., Girshick, R., He, K.: Improved baselines with momentum contrastive learning (2020)
8. Cuadros, J., Bresnick, G.: EyePACS: an adaptable telemedicine system for diabetic retinopathy screening. Diabetes Sci. Technol. **3**(3), 509–516 (2009)
9. van Ginneken, B., Stegmann, M., Loog, M.: Segmentation of anatomical structures in chest radiographs using supervised methods: a comparative study on a public database. Med. Image Anal. **10**(1), 19–40 (2006)
10. Haghighi, F., Hosseinzadeh Taher, M.R., Gotway, M.B., Liang, J.: DiRA: discriminative, restorative, and adversarial learning for self-supervised medical image analysis. In: Proceedings of the IEEE/CVF Conference on Computer Vision and Pattern Recognition (CVPR), pp. 20824–20834 (2022)
11. Haghighi, F., Taher, M.R.H., Zhou, Z., Gotway, M.B., Liang, J.: Transferable visual words: exploiting the semantics of anatomical patterns for self-supervised learning. IEEE Trans. Med. Imaging **40**(10), 2857–2868 (2021)
12. Hosseinzadeh Taher, M.R., Haghighi, F., Feng, R., Gotway, M.B., Liang, J.: A systematic benchmarking analysis of transfer learning for medical image analysis. In: Domain Adaptation and Representation Transfer, and Affordable Healthcare and AI for Resource Diverse Global Health, pp. 3–13 (2021)
13. Hosseinzadeh Taher, M.R., Haghighi, F., Gotway, M.B., Liang, J.: CAiD: context-aware instance discrimination for self-supervised learning in medical imaging. In: Proceedings of The 5th International Conference on Medical Imaging with Deep Learning. Proceedings of Machine Learning Research, vol. 172, pp. 535–551 (2022)
14. Jaeger, S., Candemir, S., Antani, S., Wáng, Y.X.J., Lu, P.X., Thoma, G.: Two public chest x-ray datasets for computer-aided screening of pulmonary diseases. Quant. Imaging Med. Surg. **4**(6) (2014)
15. Jiang, Y., Sun, M., Guo, H., Yan, K., Lu, L., Xu, M.: Anatomical invariance modeling and semantic alignment for self-supervised learning in 3D medical image segmentation. arXiv (2023)
16. Kaku, A., Upadhya, S., Razavian, N.: Intermediate layers matter in momentum contrastive self supervised learning. In: Advances in Neural Information Processing Systems, pp. 24063–24074 (2021)

17. Lian, J., et al.: A structure-aware relation network for thoracic diseases detection and segmentation. IEEE Trans. Med. Imaging **40**(8), 2042–2052 (2021)
18. Manjoo, F.: How Do You Know a Human Wrote This. The New York Times (2020)
19. Manning, C.D., Clark, K., Hewitt, J., Khandelwal, U., Levy, O.: Emergent linguistic structure in artificial neural networks trained by self-supervision. Proc. Natl. Acad. Sci. **117**(48), 30046–30054 (2020)
20. Nguyen, H.Q., Lam, K., Le, L.T., et al.: VinDr-CXR: an open dataset of chest x-rays with radiologist's annotations. Sci. Data **9**, 429 (2020)
21. Nguyen, H.C., Le, T.T., Pham, H.H., Nguyen, H.Q.: VinDr-RibCXR: a benchmark dataset for automatic segmentation and labeling of individual ribs on chest X-rays. In: Medical Imaging with Deep Learning (2021)
22. OpenAI: GPT-4 technical report (2023)
23. Ramesh, A., et al.: Zero-shot text-to-image generation. In: Proceedings of the 38th International Conference on Machine Learning, vol. 139, pp. 8821–8831 (2021)
24. Sun, Y., Hu, J., Shi, J., Sun, Z.: Progressive decomposition: a method of coarse-to-fine image parsing using stacked networks. Multimedia Tools Appl. **79**(19–20), 13379–13402 (2020)
25. Tajbakhsh, N., Roth, H., Terzopoulos, D., Liang, J.: Guest editorial annotation-efficient deep learning: the holy grail of medical imaging. IEEE Trans. Med. Imaging **40**(10), 2526–2533 (2021)
26. Tang, Y., et al.: Self-supervised pre-training of swin transformers for 3D medical image analysis. In: Proceedings of the IEEE/CVF Conference on Computer Vision and Pattern Recognition (CVPR), pp. 20730–20740 (2022)
27. Wang, X., Peng, Y., Lu, L., Lu, Z., Bagheri, M., et al.: ChestX-ray8: hospital-scale chest X-ray database and benchmarks on weakly-supervised classification and localization of common thorax diseases. In: Proceedings of the IEEE/CVF Conference on Computer Vision and Pattern Recognition (CVPR), pp. 2097–2106 (2017)
28. Wang, X., Zhang, R., Shen, C., Kong, T., Li, L.: Dense contrastive learning for self-supervised visual pre-training. In: Proceedings of the IEEE/CVF Conference on Computer Vision and Pattern Recognition (CVPR), pp. 3024–3033 (2021)
29. Xiao, J., Bai, Y., Yuille, A., Zhou, Z.: Delving into masked autoencoders for multi-label thorax disease classification. In: Proceedings of the IEEE/CVF Winter Conference on Applications of Computer Vision (WACV), pp. 3588–3600 (2023)
30. Xie, Z., et al.: SimMIM: a simple framework for masked image modeling. In: Proceedings of the IEEE/CVF Conference on Computer Vision and Pattern Recognition, pp. 9653–9663 (2022)
31. Yan, K., et al.: SAM: self-supervised learning of pixel-wise anatomical embeddings in radiological images. IEEE Trans. Med. Imaging **41**(10), 2658–2669 (2022)
32. Zhou, H.Y., Lu, C., Yang, S., Han, X., Yu, Y.: Preservational learning improves self-supervised medical image models by reconstructing diverse contexts. In: Proceedings of the IEEE/CVF International Conference on Computer Vision (ICCV), pp. 3499–3509 (2021)

The Performance of Transferability Metrics Does Not Translate to Medical Tasks

Levy Chaves[1]([⊠]), Alceu Bissoto[1], Eduardo Valle[2,3], and Sandra Avila[1]

[1] Recod.ai Lab, Institute of Computing, University of Campinas, Campinas, Brazil
{levy.chaves,alceubissoto,sandra}@ic.unicamp.br
[2] School of Electrical and Computing Engineering, University of Campinas,
Campinas, Brazil
dovalle@dca.fee.unicamp.br
[3] Valeo.ai Paris, Paris, France

Abstract. Transfer learning boosts the performance of medical image analysis by enabling deep learning (DL) on small datasets through the knowledge acquired from large ones. As the number of DL architectures explodes, exhaustively attempting all candidates becomes unfeasible, motivating cheaper alternatives for choosing them. Transferability scoring methods emerge as an enticing solution, allowing to efficiently calculate a score that correlates with the architecture accuracy on any target dataset. However, since transferability scores have not been evaluated on medical datasets, their use in this context remains uncertain, preventing them from benefiting practitioners. We fill that gap in this work, thoroughly evaluating seven transferability scores in three medical applications, including out-of-distribution scenarios. Despite promising results in general-purpose datasets, our results show that no transferability score can reliably and consistently estimate target performance in medical contexts, inviting further work in that direction.

Keywords: Transferability Estimation · Transferability Metrics · Image Classification · Medical Applications · Transfer Learning · Deep Learning

1 Introduction

Transfer learning allows, in data-limited scenarios, to leverage knowledge obtained from larger datasets. Due to its effectiveness, it is the preferred training method in medical image analysis [14]. Practitioners typically fine-tune a pre-trained model for the target task. However, selecting the most appropriate pre-trained model can significantly impact the final performance. The growing number of architectures and datasets has led to increasingly difficult decisions. While, with unlimited resources, it would be theoretically possible to compare all options empirically, that approach is too inefficient in practice. Sound empirical

L. Koch et al. (Eds.): DART 2023, LNCS 14293, pp. 105–114, 2024.
https://doi.org/10.1007/978-3-031-45857-6_11

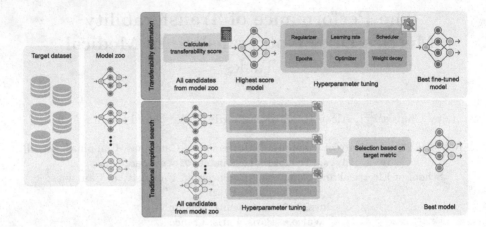

Fig. 1. Transferability estimation vs. traditional empirical search. The latter selects the best candidate model through empirical evaluation of the target metric, thus needing a costly hyperparameter search for each candidate model. Transferability computes instead a proxy score that correlates with the best expected fine-tuned performance. Only the selected model (highest score) will need hyperparameter tuning to obtain the optimized model.

evaluation must often be tempered with the designer's experience and, often, not-so-sound intuition, prejudices, and hearsay.

Transferability estimation promises to ease this burden, as shown in Fig. 1. Traditional empirical selection of architectures requires optimizing the hyperparameters of each candidate to allow a fair comparison [7]. Transferability scoring methods, in contrast, allow efficiently selecting the most promising model for a given target dataset without fine-tuning each candidate. When the transferability score accurately measures the ability to transfer knowledge between arbitrary tasks, empirical comparison of models may be limited to a small subset of candidates.

Transferability scoring methods have shown promising results, performing well when source and target datasets share strong similarities in classes and image characteristics [1,11]. However, as we will see, their behavior is much different for target medical datasets, a situation in which the target dataset deviates much more intensely from the source dataset as depicted in Fig. 2.

This work evaluates several transferability scoring methods in the medical domain, including skin lesions, brain tumors, and breast cancer. We define a comprehensive hyperparameter optimization to ensure that the fine-tuned models are evaluated on their best capabilities. Additionally, we extend the evaluation to investigate how transferability scores correlate with out-of-distribution performance. We include at least one dataset considered out-of-distribution from a source one for each medical application.

In summary, the contributions of our paper are twofold:

- We extensively evaluate seven transferability scoring methods for three distinct medical classification tasks, covering common imagery types in medical tasks;
- We design a new methodology for the medical context to account for out-of-distribution evaluation of transferability scoring methods.

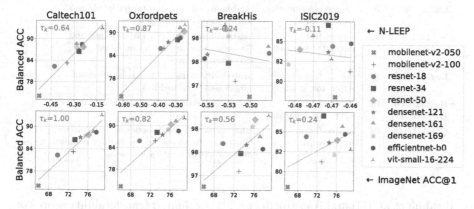

Fig. 2. Both our best transferability score (N-LEEP) and the ImageNet ACC@1 metric (baseline) on the source model suffice to predict performance on target general-purpose tasks (two left columns). For target medical tasks, the scenario is much different, with neither scores nor raw performances being strong predictors of transferability (two right columns). A good transferability scorer should capture how well a transferability score (x-axis) relates to a test performance metric (y-axis), i.e., higher values of transferability scores predict higher values of true performance. The red line showcases any linear trending between the scores and the accuracy on the task for a given source model.

2   Transferability Scores and Related Work

An effective transferability scoring method exhibits computational efficiency while strongly correlating with the final performance metric of a fine-tuned model on the target dataset. Generally, the estimation of transferability involves extracting the embeddings or predictions from the target dataset. That extracted information is integrated with the target dataset's ground-truth labels to quantify the model's transferability. Transferability scoring methods can be categorized into feature-based (fb) and source-label-based (sb). Source-label-based scores assume access to the source classification head for calculating probability distribution or label predictions, whereas feature-based scores only require source models for feature extraction. Both methods require the true labels of the target dataset for computing the transferability score. We summarize the transferability scoring methods, sorted by publication date in Table 1.

Table 1. Summary of transferability scoring methods (Tr. scorer), sorted by publication date. Cat.: category; fb: feature-based; lb: label-based.

Tr. scorer	Cat	Scorer input	Details
H-Score [2]	fb	source feature extractor & labels	transferability correlates to inter-class variance and feature redundancy
NCE [22]	lb	source classification head & labels	negative conditional entropy between source and target labels
LEEP [15]	lb	source classification head & labels	log-likelihood between target labels and source model predictions
N-LEEP [13]	fb	source feature extractor & labels	log-likelihood between target labels and Gaussian mixture model fit to target extracted features
LogME [24]	fb	source feature extractor & labels	probability of target labels conditioned on target image embeddings
Regularized H-Score [11]	fb	source feature extractor & labels	shrinkage estimators for stable covariance
GBC [17]	fb	source feature extractor & labels	Bhattacharyya coeff. between multivariate Gaussians fit to each class' feature estimating overlap with target task classes

Ibrahim et al. [11] and Agostinelli et al. [1] evaluated transferability scores on general-purpose datasets for classification and segmentation tasks. Their findings suggest that these scores may be unstable, and minor variations in the experimental protocol could lead to different conclusions. N-LEEP and LogME deliver the best transferability estimation results depending on the experimental design of classification tasks. Our work focuses on classification tasks in scenarios where the dataset shift is significant. The experimental design of previous works assumes a lower dataset shift compared to what we consider in our paper. For instance, transferring from ImageNet to CIFAR is expected to be easier than any medical dataset due to some overlap between target-source classes and features. Additionally, we perform thorough hyperparameter tuning, which is essential in these works.

3   Materials and Methods

3.1   Datasets

We assess three medical classification problems. We use ISIC2019 [5] for melanoma vs. benign classification task and PAD-UFES-20 [16] for out-of-distribution (OOD) evaluation. BreakHis [20] is used for histopathology breast cancer malign vs. benign sample classification and ICIAR2018 [18] for out-of-distribution assessment. For brain tumor classification, we use BrainTumor-Cheng [4], a four-class dataset of MRI images. We adopt the NINS [3] as the out-of-distribution test dataset.

3.2   Methodology

We aim to provide a fair and concise evaluation of each transferability scoring method described in Sect. 2. We restrict our analysis to pre-trained models on the ImageNet dataset. We focus exclusively on the *source model selection* scenario, which aims to identify the most suitable pre-trained model for a given target dataset. Our methodology involves the following seven steps:

1. Choosing a target medical task T.
2. Selecting a pre-trained model architecture A.
3. Computing the in-distribution transferability score $S_{id}(M, T, A)$ for all transferability scoring methods M, pre-trained model A, and the training dataset of task T.
4. Performing a traditional evaluation of architecture A for the target task T, by first optimizing the model hyperparameters on T's validation dataset using the target metric to obtain the best model $A_{opt}(T)$, and then evaluating that metric $P_{id}(T, A)$ on T's in-distribution test dataset.
5. Computing the out-of-distribution transferability score $S_{ood}(M, T, A)$ for all transferability scoring methods M, the fine-tuned model $A_{opt}(T)$ obtained in the previous step, and target task T's out-of-distribution test dataset (as explained in the previous subsection).
6. Evaluating the target metric $P_{ood}(T, A)$ of $A_{opt}(T)$ on T's out-of-distribution test dataset.
7. For a given dataset T and scoring method M, once steps 1–6 have been performed for all architectures, we may compute the correlation index between the transferability scores $S_*(M, T, A)$ and the traditional empirical performance metrics $P_*(T, A)$ across all architectures A. We showcase each one of those correlation analyses on a separate subplot of our results.

In our experiments, the target metric is always the balanced accuracy, and the correlation index is always the Kendall's tau, which ranges between -1 and 1, with positive correlations indicating higher-quality scoring methods. Zero correlations indicate that the scoring method has no ability to predict transferability. Negative correlations are harder to interpret: although they suggest predictive ability, they show the scoring method is working *against* its expected design.

We analyze separately the in-distribution and the out-of-distribution analyses. As far as we know, we are the first to attempt OOD analyses on transferability metrics.

4   Results

Models Architectures and Hyperparameter Tuning. We use 10 ImageNet pre-trained models: ResNet18 [9], ResNet34 [9], ResNet50 [9], MobileNetV2-0.5 [19], MobileNetV2-1.0 [19], DenseNet121 [10], DenseNet161 [10], DenseNet169 [10], EfficientNet-B0 [21], ViT-Small [6].

For hyperparameter tuning, we followed the Tuning Playbook [7] guidelines, using Halton sequences [8] to sample candidates for the hyperparameters of interest. In our search, we keep fixed the use of SGD as the optimizer, cosine scheduler, 100 epochs, and batch size of 128. We search over 75 quasi-random combinations of learning rate in range $[10^{-4}, 10^{-1}]$ and weight decay in range $[10^{-6}, 10^{-4}]$ for each model architecture, as those are the two most critical optimization hyperparameters [12]. We run the experiments on NVIDIA RTX 5000, RTX 8000. We select the best-performing model in the validation set for each architecture for test evaluation. In total, we trained 2250 models. The source code to reproduce our experiments is available at https://github.com/VirtualSpaceman/transfer-estimation-medical.

In-distribution. Figure 3 shows the results for each transferability scoring method and each model's architecture for all medical tasks. The red line indicates a regression line to show any tendency in the results. Table 2 shows all investigated transferability scores for brain tumor, histopathologic, and skin lesion classification tasks, respectively. Each row depicts one correlation index value for that transferability scoring methods (columns). We calculate each correlation index considering the test performance of the best-fine-tuned model and the estimated transferability score for each architecture.

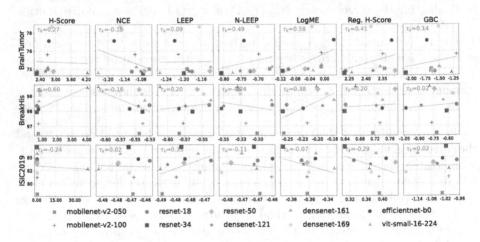

Fig. 3. Evaluation of all scores (columns) and medical datasets (rows), showcasing the correlation between transferability scores (x-axis) and best accuracy on test (y-axis). The linear regression lines (in red) are for illustration only, as the correlation index employed is the non necessarily linear Kendall's tau, shown inside of each plot, on the top-left corner. (Color figure online)

Methods such as LogME, N-LEEP, and NCE demonstrate varying degrees of correlation, indicating their potential as indicators of transferability within the same distribution. All transferability scoring methods exhibited an unstable behavior, as the correlation index consistently shifted across datasets. While

LogME was one of the best methods for the BrainTumor-Cheng and BreakHis datasets, it exhibited negative correlations for ISIC2019. To our knowledge, such instability has not been observed in general-purpose computer vision datasets. Our main hypothesis for this phenomenon relates to the dissimilarity between source and target domains. Unlike general-purpose computer vision datasets, which often overlap in target and label sets or share similar features, medical transfer learning involves substantial domain differences.

Out-of-Distribution. It is easy to come across out-of-distribution sets in the real world, as deep learning datasets often exhibit diversity and correlation shifts [23]. We conducted additional experiments to evaluate transferability scores' ability to predict out-of-distribution performance.

Table 2 shows the transferability scores and correlation indexes, with interestingly high predictive capabilities observed for label-based transferability scoring methods. NCE and LEEP exhibited outstanding results for both ICIAR2018 and PAD-UFES-20 datasets across all correlations, with NCE being favored over other methods. We hypothesize that label-based methods are prone to provide better results than feature-based for binary tasks in out-of-distribution scenarios. As the number of classes of source dataset matches the target one, the probabilities distributions tend to concentrate on a single class, inflating the transferability score for binary cases.

Table 2. Kendall's tau (τ_w) correlation index for each transferability scorer considering in and out-of-distribution scenarios for each medical task.

Task	Dataset	H-Score	NCE	LEEP	N-LEEP	LogME	Reg. H-Score	GBC
Brain Tumor	BrainTumor-cheng	0.270	−0.180	0.090	0.494	0.584	0.405	0.135
	NINS	−0.333	0.156	0.200	−0.333	−0.289	−0.422	0.200
Histopathologic	BreakHis	0.600	−0.156	0.200	−0.244	0.378	0.200	0.022
	ICIAR2018	0.333	0.778	0.778	0.289	0.289	0.378	0.156
Skin Lesion	ISIC2019	−0.244	0.022	0.333	−0.111	−0.067	−0.289	0.022
	PAD-UFES-20	−0.156	0.911	0.422	−0.156	−0.022	−0.022	0.067

Hypotheses Why Metrics Failed. Up to this point, our experiments revealed that all transferability scoring methods present unstable quality. For example, both NCE and LEEP excel at out-of-distribution but report poor results in in-distribution scenarios. We hypothesize two factors that may contribute to the failure of the methods followed by some preliminary experiments: 1) domain shift: the domain difference between source and target datasets might cause the failure. We fine-tuned each model on each medical dataset and evaluated their transferability score to the validation set. Our experiment indicates that only label-based methods excel in this scenario. So, domain shift helps to degrade the efficiency of such scores, but it is not the main reason. 2) number of classes: to measure the impact of the number of classes in the transferability scores, we take the OxfordPets dataset and map the original 37 classes (dogs and cats breeds) into a binary problem (cats vs. dogs). Our preliminary results suggest that all correlation indexes decrease, but all metrics still present high transferability estimation capabilities.

5   Conclusion

Our work is the first to investigate the quality of transferability scoring methods for medical applications. We evaluated 7 different transferability scoring methods in 3 medical classification datasets, considering 10 different architectures. Despite promising results in our out-of-distribution experiment, the instability presented by the scores across datasets in the in-distribution scenario lead us to recommend to practitioners not yet to rely on transferability scores for source model selection in medical image analysis. Our work takes one step towards reducing the need for expensive training by selecting pre-trained models efficiently that empowers the final performance on the target task. Such efficiency positively diminishes the carbon footprint when performing a hyperparameter search using a subset of deep learning architectures instead of all available.

Label-based methods shows superior results in out-of-distribution scenarios. Out-of-distribution scores might be inflated for binary tasks due to the distribution concentration on a single class, and the low number of classes benefits in favor of high transferability scores. Such an issue is absent in the available benchmarks because the general-purpose classification datasets present many classes and consider transferring from ImageNet as standard practice.

For future work, the analysis can be expanded to other configurations, such as finding the most related target task for a given source model (target model's selection) or cross-dataset transfer evaluation. Finally, evaluating future transferability scorers should include contexts where the difference between source and target domains is high, such as medical. This brings opportunities to assess the robustness of transferability scores regarding a limited amount of samples, unbalanced labels, and low inter- and high intra-variation classes.

Data Use. We use only publicity available medical datasets, including PAD-UFES-20 [16], ICIAR2018 [18], BreakHis [20], BrainTumor-Cheng [4], NINS [3], and ISIC2019 [5]. All of them are under CC BY 4.0 license, except ISIC2019 (CC BY-NC 4.0). The data collection process is described in the original papers.

Acknowledgments. L. Chaves is funded by Becas Santander/Unicamp - HUB 2022, Google LARA 2021, in part by the Coordenação de Aperfeiçoamento de Pessoal de Nível Superior - Brasil (CAPES) - Finance Code 001, and FAEPEX. A. Bissoto is funded by FAPESP (2019/19619-7, 2022/09606-8). S. Avila is funded by CNPq 315231/2020-3, FAPESP 2013/08293-7, 2020/09838-0, H.IAAC, Google LARA 2021 and Google AIR 2022.

References

1. Agostinelli, A., Pándy, M., Uijlings, J., Mensink, T., Ferrari, V.: How stable are transferability metrics evaluations? In: Avidan, S., Brostow, G., Cisse, M., Farinella, G.M., Hassner, T. (eds.) ECCV 2022. LNCS, vol. 13694, pp. 303–321. Springer, Cham (2022). https://doi.org/10.1007/978-3-031-19830-4_18
2. Bao, Y., Li, Y., Huang, S.L., Zhang, L., Zheng, L., Zamir, A., Guibas, L.: An information-theoretic approach to transferability in task transfer learning. In: International Conference on Image Processing (2019)
3. Brima, Y., Tushar, M.H.K., Kabir, U., Islam, T.: Deep transfer learning for brain magnetic resonance image multi-class classification. arXiv preprint arXiv:2106.07333 (2021)
4. Cheng, J., et al.: Enhanced performance of brain tumor classification via tumor region augmentation and partition. PLoS ONE **10**, e0140381 (2015)
5. Codella, N.C., et al.: Skin lesion analysis toward melanoma detection: a challenge at the 2017 international symposium on biomedical imaging (ISBI), hosted by the international skin imaging collaboration (ISIC). In: International Symposium on Biomedical Imaging (2018)
6. Dosovitskiy, A., et al.: An image is worth 16x16 words: transformers for image recognition at scale. In: International Conference on Learning Representations (2020)
7. Godbole, V., Dahl, G.E., Gilmer, J., Shallue, C.J., Nado, Z.: Deep learning tuning playbook (2023). version 1.0
8. Halton, J.H.: Algorithm 247: radical-inverse quasi-random point sequence. Commun. ACM **7**, 701–702 (1964)
9. He, K., Zhang, X., Ren, S., Sun, J.: Deep residual learning for image recognition. In: Computer Vision and Pattern Recognition (2016)
10. Huang, G., Liu, Z., Van Der Maaten, L., Weinberger, K.Q.: Densely connected convolutional networks. In: Computer Vision and Pattern Recognition (2017)
11. Ibrahim, S., Ponomareva, N., Mazumder, R.: Newer is not always better: rethinking transferability metrics, their peculiarities, stability and performance. In: Amini, M.R., et al. (eds.) ECML PKDD 2022. LNCS, vol. 13713, pp. 693–709. Springer, Cham (2022). https://doi.org/10.1007/978-3-031-26387-3_42
12. Li, H., et al.: Rethinking the hyperparameters for fine-tuning. In: International Conference on Learning Representations (2020)
13. Li, Y., et al.: Ranking neural checkpoints. In: Computer Vision and Pattern Recognition (2021)
14. Matsoukas, C., Haslum, J.F., Sorkhei, M., Söderberg, M., Smith, K.: What makes transfer learning work for medical images: feature reuse & other factors. In: Conference on Computer Vision and Pattern Recognition (2022)
15. Nguyen, C., Hassner, T., Seeger, M., Archambeau, C.: Leep: a new measure to evaluate transferability of learned representations. In: International Conference on Machine Learning (2020)
16. Pacheco, A.G., et al.: Pad-ufes-20: a skin lesion dataset composed of patient data and clinical images collected from smartphones. Data in Brief (2020)
17. Pándy, M., Agostinelli, A., Uijlings, J., Ferrari, V., Mensink, T.: Transferability estimation using bhattacharyya class separability. In: Computer Vision and Pattern Recognition (2022)
18. Rakhlin, A., Shvets, A., Iglovikov, V., Kalinin, A.A.: Deep convolutional neural networks for breast cancer histology image analysis. In: 15th International Conference on Image Analysis and Recognition (2018)

19. Sandler, M., Howard, A., Zhu, M., Zhmoginov, A., Chen, L.C.: Mobilenetv 2: inverted residuals and linear bottlenecks. In: Computer Vision and Pattern Recognition (2018)
20. Spanhol, F.A., Oliveira, L.S., Petitjean, C., Heutte, L.: A dataset for breast cancer histopathological image classification. IEEE Trans. Biomed. Eng. **63**, 1455–1462 (2016)
21. Tan, M., Le, Q.: Efficientnet: rethinking model scaling for convolutional neural networks. In: International Conference on Machine Learning (2019)
22. Tran, A.T., Nguyen, C.V., Hassner, T.: Transferability and hardness of supervised classification tasks. In: International Conference on Computer Vision (2019)
23. Ye, N., et al.: Ood-bench: quantifying and understanding two dimensions of out-of-distribution generalization. In: Computer Vision and Pattern Recognition (2022)
24. You, K., Liu, Y., Wang, J., Long, M.: Logme: practical assessment of pre-trained models for transfer learning. In: International Conference on Machine Learning (2021)

DGM-DR: Domain Generalization with Mutual Information Regularized Diabetic Retinopathy Classification

Aleksandr Matsun⬚, Dana O. Mohamed(✉)⬚, Sharon Chokuwa⬚, Muhammad Ridzuan⬚, and Mohammad Yaqub⬚

Mohamed Bin Zayed University of Artificial Intelligence, Abu Dhabi, UAE
{aleksandr.matsun,dana.mohamed,sharon.chokuwa,muhammad.ridzuan,
mohammad.yaqub}@mbzuai.ac.ae

Abstract. The domain shift between training and testing data presents a significant challenge for training generalizable deep learning models. As a consequence, the performance of models trained with the independent and identically distributed (i.i.d) assumption deteriorates when deployed in the real world. This problem is exacerbated in the medical imaging context due to variations in data acquisition across clinical centers, medical apparatus, and patients. Domain generalization (DG) aims to address this problem by learning a model that generalizes well to any unseen target domain. Many domain generalization techniques were unsuccessful in learning domain-invariant representations due to the large domain shift. Furthermore, multiple tasks in medical imaging are not yet extensively studied in existing literature when it comes to DG point of view. In this paper, we introduce a DG method that re-establishes the model objective function as a maximization of mutual information with a large pretrained model to the medical imaging field. We re-visit the problem of DG in Diabetic Retinopathy (DR) classification to establish a clear benchmark with a correct model selection strategy and to achieve robust domain-invariant representation for an improved generalization. Moreover, we conduct extensive experiments on public datasets to show that our proposed method consistently outperforms the previous state-of-the-art by a margin of 5.25% in average accuracy and a lower standard deviation. Source code available at https://github.com/BioMedIA-MBZUAI/DGM-DR.

Keywords: Domain Generalization · Diabetic Retinopathy · Mutual Information Regularization

A. Matsun, D. O. Mohamed and S. Chokuwa—These authors contributed equally to this work.

Supplementary Information The online version contains supplementary material available at https://doi.org/10.1007/978-3-031-45857-6_12.

L. Koch et al. (Eds.): DART 2023, LNCS 14293, pp. 115–125, 2024.
https://doi.org/10.1007/978-3-031-45857-6_12

1   Introduction

Medical imaging has become an indispensable tool in diagnosis, treatment planning, and prognosis. Coupled with the introduction of deep learning, medical imaging has witnessed tremendous progress in recent years. Notwithstanding, a major challenge in the medical imaging field is the domain shift problem, where the performance of a trained model deteriorates when for instance tested on a dataset that was acquired from a different device or patient population than the original dataset. This problem is especially prominent in tasks, where acquiring large-scale annotated datasets from one center is costly and time-consuming. Domain generalization (DG) [29] aims to alleviate this challenge by training models that can generalize well to new unseen domains, without the need for extensive domain-specific data collection and annotation.

DG in medical image analysis still requires extensive research, however there already exist a handful of works examining it. One of those works includes utilizing an adversarial domain synthesizer to create artificial domains using only one source domain to improve the generalizability of the model in downstream tasks [26]. Although such method can synthesize a wide range of possible domains, it usually suffers from the ability to mimic realistic domain shifts. Another method is applying test-time augmentations such that the target image resembles the source domain, thus reducing the domain shift and improving generalization [25]. Moreover, DRGen [3] combines Fishr [20] and Stochastic Weight Averaging Densely (SWAD) [6] to achieve domain generalization in Diabetic Retinopathy (DR) classification. In DRGen, Fishr [20] is used to make the model more robust to variations in the data by penalizing large differences in the gradient variances between in-distribution and out-of-distribution data, and SWAD [6] is used to seek flatter minima in the loss landscape of the model. DRGen is currently state-of-the-art in DR classification, however it has been evaluated using samples from the testing set which makes it harder to assess its true generalizability.

In natural images, the domain generalization problem has been explored extensively compared to medical imaging analysis. Some of the DG methods proposed over the past ten years include domain alignment [16], meta-learning [10], style transfer [28], and regularization methods [14]. More recently, the authors of [7] utilize a large pretrained model to guide a target model towards generalized feature representation through mutual information regularization. Another DG regularization method that can be applied orthogonally to many DG algorithms is SWAD [6], which improves domain generalizability by seeking flat minima in the loss landscape of the model. The flatter minima indicate that the loss is not changing significantly in any direction, thus reducing the risk of the model overfitting to domain biases [6]. However, when adapting a DG approach that demonstrates a good performance on natural images, there is no guarantee of a similar performance on medical imaging applications due to the typical complex nature of such problems.

DR is a complication of Diabetes Mellitus that affects the eyes and can lead to vision loss or blindness. It is caused by damage to the blood vessels in the retina due to high blood sugar levels, which often leads to blood leakage onto the retina

[24]. This can cause swelling and distortion of vision. The prevalence of DR is increasing worldwide due to the growing number of people with diabetes. However, early detection and management of DR is critical to the prevention of vision deterioration or loss. DR can be classified into 4 classes: mild, moderate, severe, and proliferative . Some of the visible features that are used to classify the first 3 classes include microaneurysms, retinal hemorrhages, intraretinal microvascular abnormalities (IRMA), and venous caliber changes, while pathologic preretinal neovascularization is used to classify proliferative DR [9].

In this paper, we propose DGM-DR, a Domain Generalization with Mutual information regularized Diabetic Retinopathy classifier. Our main contributions are as follows:

- We introduce a DG method that utilizes mutual information regularization with a large pretrained oracle model.
- We show the improvement of our proposed solution on the DR classification task over the previous state-of-the-art in both performance and robustness through rigorous investigations.
- We set a clear benchmark with the correct DG model selection method inline with standard DG protocols for the task of DR classification.

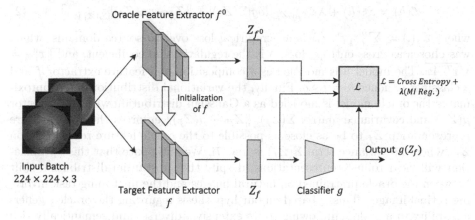

Fig. 1. Overview of the proposed method, DGM-DR. It consists of the oracle f^0 and target f feature extractor, where f is initialized by the weights of f^0. For each sampled mini-batch, feature representations are extracted using both feature extractors f^0 and f. Features Z_f are then passed to classifier g. Lastly, the loss - a linear combination of cross entropy and mutual information regularization loss - is calculated, and f and g are updated.

2   Methodology

Our work is inspired by [7], which aims to improve model generalizability when classifying natural images. In DGM-DR, we re-establish the domain generalization objective as a maximization of mutual information with a large pretrained model, named the oracle, to address DR classification. We aim to make the distribution of feature representations of the target model close to the generalized one of the oracle by maximizing the mutual information between both. The oracle model is trained on a large-scale diverse dataset that contains information on many different domains in order to approximate it as closely as possible to a true oracle, which is a model that can generalize to any domain and is inaccessible in practice. Figure 1 shows an overview of DGM-DR's process. Initially, the oracle's weights are used to initialize the target model's feature extractor. Then, for each mini-batch, the oracle feature extractor f^0 and the target feature extractor f are used to extract feature representations Z_{f^0} and Z_f, respectively. The features Z_f are passed to the classifier g to produce the output. The oracle model is chosen as ImageNet pretrained ResNet-50 [13] for a realistic and fair comparison with other DG algorithms. It is shown in [5] that maximization of the lower bound of the mutual information between Z_{f^0} and Z_f is equivalent to minization of the term 1

$$\mathbb{E}_{Z_{f^0},Z_f}\left[log|\Sigma(Z_f)| + \|Z_{f^0} - \mu(Z_f)\|^2_{\Sigma(Z_f)^{-1}}\right] \tag{1}$$

The final loss is calculated using Eq. 2:

$$\mathcal{L}(h) = \mathcal{E}_S(h) + \lambda\mathbb{E}_{Z_{f^0},Z_f}\left[log|\Sigma(Z_f)| + \|Z_{f^0} - \mu(Z_f)\|^2_{\Sigma(Z_f)^{-1}}\right] \tag{2}$$

where $\mathcal{E}_S(.) = \sum_{d=1}^{m} \mathcal{E}_{S_d}(.)$ is an empirical loss over m source domains, which was chosen as cross-entropy loss, λ is the regularization coefficient, and $\|x\|_A = \sqrt{x^T A x}$. The model h is modeled as a composition of a feature extractor f and a classifier g, hence $h = f \circ g$. Finally, the variational distribution that approximates the oracle model is modeled as a Gaussian distribution with mean vector $\mu(Z_f)$ and covariance matrix $\Sigma(Z_f)$. $\|Z_{f^0} - \mu(Z_f)\|$ enforces the mean feature representation Z_f to be as close as possible to the oracle feature representation Z_{f^0} when the variance term $\Sigma(Z_f)$ is low [7]. We anticipate that this optimization will yield robust representations, despite the substantial distribution shift between the oracle pretrained on natural images and the finetuning task involving retinal images. This is based on our hypothesis regarding the oracle's generalizability to any domain, owing to its extensive, diverse, and semantically rich features that surpass those found in any other medical dataset. The regularization term λ aims to minimize the variance in the target features and encourage similarity between the oracle and target features. This, in turn, facilitates the learning of domain-invariant representations that generalize well across different domains.

3   Experimental Setup

3.1   Datasets

We utilize the four datasets used by [3], which are EyePACS [2], APTOS [1], Messidor and Messidor-2 [17]. The 4 datasets are composed of 5 classes of 5 grades from 0 to 4: No DR (Grade 0), mild DR (Grade 1), moderate DR (Grade 2), severe DR (Grade 3), and proliferative DR (Grade 4). These datasets were acquired from various geographical regions, encompassing India, America, and France [1,2,17]. As a result, domain shift emerges, due to the variations in the employed cameras [2,4], and the difference in population groups. Figure 2 shows example images for the 5 DR classes. A breakdown of the distribution of the classes is given in Table A.4. In all 4 datasets, there is a high imbalance between the no DR class and the other 4 DR classes.

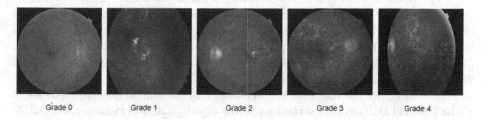

Grade 0   Grade 1   Grade 2   Grade 3   Grade 4

Fig. 2. Sample images from different DR classes obtained from APTOS [1].

3.2 Data Augmentations

All fundus images are resized to $224 \times 224 \times 3$. We perform histogram equalization with a probability $p = 0.5$, horizontal flip and color jitter by a value of 0.3 in brightness, contrast, saturation, and hue with $p = 0.3$.

3.3 Evaluation Methods

We utilize the DomainBed [11] evaluation protocols for fair comparison with DRGen [3] and other DG algorithms. The appropriate DG model selection method used is the training-domain validation set following DomainBed [11], in which we split each training domain into training and validation subsets, pool the validation subsets together to create an overall validation set, and finally choose the model that maximizes the accuracy on the overall validation set. We use 20% of the source training data for validation. We evaluate the performance scores using leave-one-domain-out cross validation, and average the cases where a specific domain is used as a target domain and the others as source domains.

We also perform comparisons of the proposed and existing DG approaches with the Empirical Risk Minimization (ERM) technique that aims to minimize in-domain errors. Interestingly, [11] argues that carefully training a model using ERM achieves a near state-of-the-art performance. This was tested on a range of baselines and was shown to outperform a few DG models.

3.4   Implementation Details

We implement all our models using the PyTorch v1.7 framework. The experiments were run on 24GB Quadro RTX 6000 GPU. The backbone used is ResNet-50 pretrained on ImageNet. We use the Adam optimizer [15] with a learning rate of $5e - 5$ and no weight decay, chosen experimentally. The model was trained in 5000 steps. The batch size was fixed to 32 images. The λ regularization coefficient was set to 1.0. Different values of lambda were experimented with, and the results are given in A.5.

To compare against other DG methods, we reproduce the results of all algorithms using the same implementation details mentioned previously for a fair comparison. For Fishr [20], we set the Fishr lambda (λ) to 1000, penalty anneal iteration (γ) of 1500 and an exponential moving average of 0.95. For DRGen [3], we use SWAD as the model selection method as opposed to the test-domain validation used in the original paper [3], which is not suitable for DG evaluation. Moreover, we use the data augmentations in the official implementations of Fishr [20] and DRGen [3] for the respective algorithms, otherwise we use DGM-DR's augmentations. Finally, we use SWAD as the model selection method when combining DGM-DR with the SWAD [6] algorithm.

Table 1. Multi-class classification results with ERM and DG methods averaged over three runs. The best accuracy (%) is highlighted in bold.

Algorithm	APTOS	EyePACS	Messidor	Messidor-2	Average Accuracy
ERM [22]	62.83	73.01	**66.88**	65.26	66.99±4.3
Fishr [20]	56.49	68.24	61.53	62.11	62.09 ±4.8
DRGen [3]	54.53	**73.87**	52.03	69.13	62.39±9.3
DGM-DR	**65.39**	70.12	65.63	**69.41**	67.64±2.4
DGM-DR + SWAD [6]	65.15	71.92	65.66	68.96	**67.92±3.2**

4   Results

Table 1 compares the performance of DGM-DR with three other methods, including the previous state-of-the-art DRGen. The experiments for the main results were repeated three times using three different seeds, and the average accuracy and standard deviation across the runs are reported. DGM-DR achieves > 5% increase in the average accuracy when compared with the other DG methods (Fishr and DRGen) and 1% increase compared to ERM-based model [22].

4.1   Ablation Studies

Changing the Oracle Pretraining Datasets, Methods, and Backbones.
We investigate the effect of changing the oracle on the DR classification task and report the results in Table 2. We use ImageNet pretrained ResNet-50 using Barlow Twins [27] and MoCo [12], CLIP pretrained ResNet-50, and large-scale pretraining including CLIP pretrained ViT-B [8] and SWAG pretrained

RegNetY-16GF [19]. All experiments were performed with the same implementation details mentioned previously, except for RegNetY-16GF, where the batch size was changed from 32 to 16 due to hardware limitations.

Table 2. Results of changing the oracle pretraining datasets, methods, and backbones. SWAG* is using a batch size of 16, while the rest of the experiments are using a batch size of 32. The average accuracy and the standard deviation across the 4 domains in a single run are given, with the best accuracy (%) highlighted in bold.

Dataset	Pre-training	APTOS	EyePACS	Messidor	Messidor 2	Average Accuracy
ImageNet	ERM	**65.39**	70.12	65.63	**69.41**	**67.64±2.5**
	Barlow Twins	60.66	73.45	55.57	61.18	62.71±7.6
	MoCo v3	56.90	72.69	65.77	68.41	65.94±6.7
CLIP	CLIP (ResNet)	61.01	73.33	62.44	58.10	63.72±6.7
	CLIP (ViT)	64.25	68.54	**66.29**	66.05	66.28±1.8
Instagram	SWAG* (RegNet)	63.12	**75.38**	62.96	64.61	66.52±6.0

Binary Classification of DR. We study the effect of changing the multi-class classification task into a binary classification task, where fundus images are classified as *DR* or *No DR*. The results of this experiment are reported in Table 3.

5   Discussion

In Table 1, we report the results of 4 different algorithms and show that DGM-DR outperforms all algorithms, including the previous state-of-the-art DRGen [3] by a significant margin of 5.25%. Additionally, DGM-DR demonstrates robustness with a relatively small standard deviation of 2.4 across three different experiments. As was concluded in [11], ERM-based methods can outperform a range of DG methods, if carefully trained. We show in Table 1 that the ERM method outperforms existing DG baselines that we compare with. On the other hand, we show that DGM-DR outperforms the ERM's performance for multiclass classification. We believe that even though the task of DR classification is challenging, the fundus images across all domains share common semantic structures, hence ERM is able to learn some domain-invariant features. However, the performance of DGM-DR is more stable, with a standard deviation being almost half that of ERM's. This can be attributed to DGM-DR's novel learning technique that aims to minimize a combination of cross entropy and mutual information regularization with an oracle, which enables it to learn more robust domain-invariant representations. Lastly, with the addition of SWAD to DGM-DR, performance further increases by a slight value (0.28%), consistent with previous literature (e.g. [21]) where accuracy is improved when combined with SWAD.

Table 3. Results of the binary classification task using different algorithms. The average accuracy and the standard deviation across the 4 domains in a single run are given, with the best accuracy (%) highlighted in bold.

Algorithm	APTOS	EyePACS	Messidor	Messidor-2	Average Accuracy
ERM	**95.42**	74.70	**86.98**	77.47	**83.63±9.5**
Fishr	90.67	74.45	77.92	**79.30**	80.59±7.0
DRGen	82.05	**75.41**	81.67	72.42	77.89±4.7
DGM-DR	88.34	71.82	86.15	78.10	80.00±8.2
DGM-DR + SWAD	88.00	72.06	85.63	76.22	80.48±7.6

In general, the performance of all algorithms on each of the datasets is consistent. This indicates that any decline or increase in performance of a dataset can be attributed to the distribution of the dataset itself, which is used as the target domain in the evaluation, and to the distribution of the combined source datasets on which the model is trained. For example, EyePACS [2] consistently performs better across all algorithms. A possible yet arguable hypothesis is it is highly imbalanced, as demonstrated in Table A.4, with the majority of images belonging to *No DR*. Since the *No DR* class is the majority in all datasets, the model is also biased towards it. Hence the model could be correctly classifying the *No DR* case and randomly guessing in the other four cases.

In Table 2, we study the effect of changing the oracle pretraining datasets, methods, and backbones. The large SWAG* pretrained RegNetY-16GF oracle yields the best accuracy in this experiment, second only to our ResNet-50 with ImageNet ERM pretraining, possibly due to the smaller batch size and the limit of number of steps set for a fair comparison. In general, we observe that a larger oracle model trained on a bigger, more diverse dataset is able to guide the target model towards more generalized feature representations. However, it will require longer training time to converge.

In Table 3, we notice that ERM is doing a better job at binary classification of DR than DGM-DR. Since the binary classification problem is simpler, as is visually evident in Fig. 2, DG algorithms tend to negatively impact the results as they are likely to introduce more complexity. Furthermore, the generalization gap [23] is typically smaller in binary classification than a multiclass setup. Therefore, ERM-based methods are likely to outperform DG-based methods in such scenarios.

The selection of the mutual information regularization coefficient λ, which controls the balance between the cross entropy loss and the mutual information regularization loss, is related to how informative the oracle model's knowledge is for the target model's task. A large λ encourages the model to reduce the variance in the target features and enforce similarity between the target and oracle features. Thus, the model will focus on learning domain-invariant patterns originating from the oracle's knowledge, which is ImageNet in our main experiment. On the other hand, a small λ reduces the emphasis on domain-invariance and thus may potentially lead to overfitting.

In our case, as shown in [18], ImageNet initialization of deep learning models is beneficial in the context of medical imaging analysis, including fundus images. Therefore, we conclude that the best λ for the case of DR classification is 1.0 for the ImageNet pretrained ResNet-50, in contrast with that of natural images where λ is typically set to be $\{0.001, 0.01, 0.1\}$ as in [7]. We believe that in the DR classification problem, the oracle model has a significant impact on training the target DG model due to its rich low level feature representations which cannot be easily learnt from scratch or from a small size dataset.

As a final note, a very important part of a domain generalization solution is the model selection method, as it simplifies fair assessments by disregarding differences in results due to inconsistent hyperparameter tuning that may be attributed to the algorithms under study [11]. Furthermore, utilizing the test-domain validation set as a model selection method is inappropriate for a DG algorithm, which was done by DRGen [3] in DR classification. Hence, one goal of this paper is to set a clear benchmark for DR classification using training-domain validation, thus allowing easy comparison with future work.

6   Conclusion

In this paper, we introduce DGM-DR to tackle the problem of DR classification with domain generalization. Our use of a large pretrained model to guide the target model towards learning domain-invariant features across different DR datasets through mutual information regularization achieves superior performance over the previous-state-of-the art DG methods. We also establish a clear benchmark for the task using a DG-appropriate model selection algorithm, thus allowing future work to make comparisons with our work. Further investigation to understand when and why DG-based methods could be superior or inferior to ERM-based approaches in medical imaging is needed. Although we believe that our work pushes the horizons of the DG field in medical image analysis, several DG-related research questions are yet to be investigated e.g., unsupervised DG, interpretable DG, and performance evaluation to DG methods.

References

1. Aptos 2019 Blindness Detection. https://www.kaggle.com/c/aptos2019-blindness-detection/data
2. Kaggle: Diabetic Retinopathy Detection - EYEPACS Dataset. https://www.kaggle.com/c/diabetic-retinopathy-detection
3. Atwany, M., Yaqub, M.: Drgen: domain generalization in diabetic retinopathy classification. In: Wang, L., Dou, Q., Fletcher, P.T., Speidel, S., Li, S. (eds.) Medical Image Computing and Computer Assisted Intervention - MICCAI 2022, pp. 635–644. Springer, Cham (2022)
4. Atwany, M.Z., Sahyoun, A.H., Yaqub, M.: Deep learning techniques for diabetic retinopathy classification: a survey. IEEE Access (2022)
5. Barber, D., Agakov, F.: The im algorithm: a variational approach to information maximization. Adv. Neural. Inf. Process. Syst. **16**(320), 201 (2004)

6. Cha, J., et al.: Swad: domain generalization by seeking flat minima (2021). https:// doi.org/10.48550/ARXIV.2102.08604. https://arxiv.org/abs/2102.08604

7. Cha, J., Lee, K., Park, S., Chun, S.: Domain generalization by mutual-information regularization with pre-trained models (2022). https://doi.org/10.48550/ARXIV. 2203.10789. https://arxiv.org/abs/2203.10789

8. Dosovitskiy, A., et al.: An image is worth 16×16 words: transformers for image recognition at scale (2020). https://doi.org/10.48550/ARXIV.2010.11929. https:// arxiv.org/abs/2010.11929

9. Duh, E.J., Sun, J.K., Stitt, A.W.: Diabetic retinopathy: current understanding, mechanisms, and treatment strategies. JCI Insight **2**(14), July 2017

10. Finn, C., Abbeel, P., Levine, S.: Model-agnostic meta-learning for fast adaptation of deep networks. CoRR abs/1703.03400 (2017). http://arxiv.org/abs/1703.03400

11. Gulrajani, I., Lopez-Paz, D.: In search of lost domain generalization (2020). https://doi.org/10.48550/ARXIV.2007.01434

12. He, K., Fan, H., Wu, Y., Xie, S., Girshick, R.: Momentum contrast for unsupervised visual representation learning (2019). https://doi.org/10.48550/ARXIV. 1911.05722. https://arxiv.org/abs/1911.05722

13. He, K., Zhang, X., Ren, S., Sun, J.: Deep residual learning for image recognition (2015). https://doi.org/10.48550/ARXIV.1512.03385. https://arxiv.org/abs/1512. 03385

14. Huang, Z., Wang, H., Xing, E.P., Huang, D.: Self-challenging improves cross-domain generalization. CoRR abs/2007.02454 (2020). https://arxiv.org/abs/2007. 02454

15. Kingma, D.P., Ba, J.: Adam: A method for stochastic optimization (2014). https:// doi.org/10.48550/ARXIV.1412.6980. https://arxiv.org/abs/1412.6980

16. Li, H., Pan, S.J., Wang, S., Kot, A.C.: Domain generalization with adversarial feature learning. In: 2018 IEEE/CVF Conference on Computer Vision and Pattern Recognition, pp. 5400–5409 (2018). https://doi.org/10.1109/CVPR.2018.00566

17. Maffre, Gauthier, G., Lay, B., Roger, J., Elie, D., Foltete, M., Donjon, A., Hugo, Patry, G.: Messidor. https://www.adcis.net/en/third-party/messidor/

18. Matsoukas, C., Haslum, J.F., Sorkhei, M., Söderberg, M., Smith, K.: What makes transfer learning work for medical images: feature reuse & other factors (2022). https://doi.org/10.48550/ARXIV.2203.01825. https://arxiv.org/abs/2203.01825

19. Radosavovic, I., Kosaraju, R.P., Girshick, R., He, K., Dollár, P.: Designing network design spaces (2020). https://doi.org/10.48550/ARXIV.2003.13678, https://arxiv. org/abs/2003.13678

20. Ramé, A., Dancette, C., Cord, M.: Fishr: invariant gradient variances for out-of-distribution generalization. CoRR abs/2109.02934 (2021). https://arxiv.org/abs/ 2109.02934

21. Rangwani, H., Aithal, S.K., Mishra, M., Jain, A., Radhakrishnan, V.B.: A closer look at smoothness in domain adversarial training. In: International Conference on Machine Learning, pp. 18378–18399. PMLR (2022)

22. Vapnik, V.N.: Statistical Learning Theory. Wiley, NY (1998)

23. Vedantam, R., Lopez-Paz, D., Schwab, D.J.: An empirical investigation of domain generalization with empirical risk minimizers. In: Ranzato, M., Beygelzimer, A., Dauphin, Y., Liang, P., Vaughan, J.W. (eds.) Advances in Neural Information Processing Systems, vol. 34, pp. 28131–28143. Curran Associates, Inc. (2021). https://proceedings.neurips.cc/paper/2021/file/ ecf9902e0f61677c8de25ae60b654669-Paper.pdf

24. Wang, W., Lo, A.C.Y.: Diabetic retinopathy: Pathophysiology and treatments. Int. J. Mol. Sci. **19**(6), June 2018

25. Xu, C., Wen, Z., Liu, Z., Ye, C.: Improved domain generalization for cell detection in histopathology images via test-time stain augmentation. In: Wang, L., Dou, Q., Fletcher, P.T., Speidel, S., Li, S. (eds.) Medical Image Computing and Computer Assisted Intervention - MICCAI 2022, pp. 150–159. Springer, Cham (2022)

26. Xu, Y., Xie, S., Reynolds, M., Ragoza, M., Gong, M., Batmanghelich, K.: Adversarial consistency for single domain generalization in medical image segmentation (2022). https://doi.org/10.48550/ARXIV.2206.13737. https://arxiv.org/abs/2206.13737

27. Zbontar, J., Jing, L., Misra, I., LeCun, Y., Deny, S.: Barlow twins: self-supervised learning via redundancy reduction (2021). https://doi.org/10.48550/ARXIV.2103.03230. https://arxiv.org/abs/2103.03230

28. Zhang, Y., Li, M., Li, R., Jia, K., Zhang, L.: Exact feature distribution matching for arbitrary style transfer and domain generalization (2022). https://doi.org/10.48550/ARXIV.2203.07740. https://arxiv.org/abs/2203.07740

29. Zhou, K., Liu, Z., Qiao, Y., Xiang, T., Loy, C.C.: Domain generalization: a survey. IEEE Transactions on Pattern Analysis and Machine Intelligence, pp. 1–20 (2022). https://doi.org/10.1109/tpami.2022.3195549. https://doi.org/10.1109%2Ftpami.2022.3195549

SEDA: Self-ensembling ViT with Defensive Distillation and Adversarial Training for Robust Chest X-Rays Classification

Raza Imam[✉], Ibrahim Almakky, Salma Alrashdi, Baketah Alrashdi, and Mohammad Yaqub

Mohamed Bin Zayed University of Artificial Intelligence,
Abu Dhabi, UAE
{raza.imam,ibrahim.almakky,salma.alrashdi,baketah.alrashdi,
mohammad.yaqub}@mbzuai.ac.ae

Abstract. Deep Learning methods have recently seen increased adoption in medical imaging applications. However, elevated vulnerabilities have been explored in recent Deep Learning solutions, which can hinder future adoption. Particularly, the vulnerability of Vision Transformer (ViT) to adversarial, privacy, and confidentiality attacks raise serious concerns about their reliability in medical settings. This work aims to enhance the robustness of self-ensembling ViTs for the tuberculosis chest x-ray classification task. We propose Self-Ensembling ViT with defensive Distillation and Adversarial training (SEDA). SEDA utilizes efficient CNN blocks to learn spatial features with various levels of abstraction from feature representations extracted from intermediate ViT blocks, that are largely unaffected by adversarial perturbations. Furthermore, SEDA leverages adversarial training in combination with defensive distillation for improved robustness against adversaries. Training using adversarial examples leads to better model generalizability and improves its ability to handle perturbations. Distillation using soft probabilities introduces uncertainty and variation into the output probabilities, making it more difficult for adversarial and privacy attacks. Extensive experiments performed with the proposed architecture and training paradigm on publicly available Tuberculosis x-ray dataset shows SOTA efficacy of SEDA compared to SEViT in terms of computational efficiency with 70× times lighter framework and enhanced robustness of +9%. Code: Github.

Keywords: Ensembling · Adversarial Attack · Defensive Distillation · Adversarial Training · Vision Transformer

Supplementary Information The online version contains supplementary material available at https://doi.org/10.1007/978-3-031-45857-6_13.

1   Introduction

Deep Learning (DL) models have proven their efficacy on various medical tasks in general and on classifying lung diseases from x-ray images in particular [1]. However, popular DL models, such as Convolutional Neural Networks (CNNs) and Vision Transformers (ViTs) are susceptible to adversarial attacks. The vulnerabilities of medical imaging systems to such attacks are increasingly apparent [2]. These attacks can target cloud-based processing or communication channels to compromise patient data [3]. Adversarial attacks in medical imaging can also be exploited for financial gain through fraudulent billing or insurance claims [4]. Moreover, malicious parties extracting model information could compromise patient data privacy. Therefore, it is essential to develop robust defense strategies to ensure the secure deployment of automated medical imaging systems [5]. Detecting and defending against adversarial attacks is crucial for accurate diagnoses and successful treatment outcomes in the healthcare domain. [6].

Recently, SEViT [4], a self-ensembling approach based on Vision Transformers, has shown promise in defending against adversarial attacks in medical image analysis. However, SEViT has practical deployment limitations due to the large parameter size of the MLP modules ensembled for each block. Attacks such as model extraction can extract SEViT models, despite their effectiveness against adversarial attacks [7]. Privacy attacks, such as model extraction in the medical imaging domain, pose significant threats to patient privacy and confidentiality. For instance, attackers can extract trained models from healthcare facilities and exploit them for malicious purposes, including selling the models, training their own models using the extracted models, or inferring sensitive patient information from x-ray images.

Medical data can contain sensitive information about patients, and unauthorized model extraction can lead to privacy violations [8]. To defend against such model extraction attacks, defensive distillation [9] can be employed. Defensive distillation enhances model robustness by training a distilled model on softened probabilities from an initial model, mitigating adversarial attacks by introducing prediction uncertainty, and reducing sensitivity to small perturbations [5]. Soft probabilities refer to the output of a model's prediction that represents the likelihood or probability distribution over multiple classes rather than a single deterministic label. By training a distilled model that approximates the behavior of the original model, we can enhance resistance to extraction. By making extraction more challenging, attackers may be deterred due to the increased time and resources required. In this work, we aim to improve the robustness of the SEViT model for Tuberculosis (TB) classification from chest x-ray images using defensive distillation and adversarial training, while also improving computational efficiency. The main aim of this research is to provide a lightweight, accurate, and robust benchmark medical imaging system, which aims to be crucial in achieving better diagnosis and treatment outcomes for patients and healthcare providers. The main contributions of this work are as follows:

- We propose an efficient and lightweight Self-Ensembling ViT with Defensive Distillation and Adversarial training (SEDA) to increase adversarial robustness while preserving the clean accuracy of ViT by combining the adversarial training and defensive distillation approaches.

- Showing the real-world deployability of SEDA, we evaluate the proposed framework's computational parameters and classification performance, and compare them with state-of-the-art methods [4,10].

2   Related Works

Ensembling ViTs. Adversarial attacks can be more potent against ViTs by targeting both intermediate representations and the final class token [11]. To address this challenge, [4] proposed the Self-Ensemble Vision Transformer (SEViT) architecture. SEViT enhances the robustness of ViT against adversarial attacks in medical image classification. They propose to add an MLP classifier at the end of each block to leverage patch tokens and generate a probability distribution over class labels. This approach enables the self-ensembling of classifiers, which can be fused to obtain the final classification result.

Adversarial Attacks. The works by [2,12,13] analyzed a common theme of exploring adversarial attacks in machine learning. They aim to better understand the nature of these attacks and propose new techniques for mitigating their effects on learning models. [12] presents a new technique for creating adversarial examples, while [2] proposes a defense mechanism against them. [13] presents an efficient approach to training models to resist adversarial attacks.

Adversarial Defense. [9] proposes a defense mechanism called defensive distillation, which is a form of knowledge transfer from a larger, more accurate model to a smaller, less accurate model. The idea of [9] is to make the smaller model less vulnerable to adversarial attacks by training it to imitate the outputs of the larger model, which is assumed to be more robust to adversarial perturbations.

3   Proposed Method

The objective of an adversarial attack is to generate a perturbed image x', which is similar to the original medical image x within a certain distance metric (L_∞ norm), such that the output of a ViT-based classifier $f(x')$ is different from the true label y with a high probability. Defending against adversarial attacks involves obtaining a robust classifier f' from the original classifier f. This robust classifier f' should have high accuracy on both clean images ($P(f'(x) = y)$) and perturbed images ($P(f'(x') = y)$).

Ensembling SEViT [4] aims to improve adversarial robustness by adding an MLP classifier at the end of first m blocks, utilizing patch tokens to produce a probability distribution over class labels. The intermediate feature representations output by the initial blocks are considered useful for classification and more robust against adversarial attacks. This results in a self-ensemble of L classifiers that can be fused to obtain the final classification result. The SEViT model increases adversarial robustness by adding MLP classifiers only to the

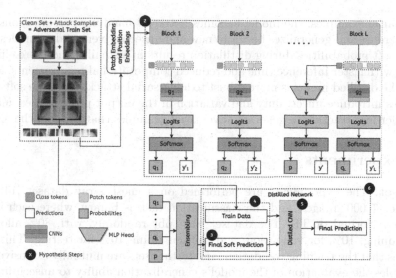

Fig. 1. The proposed SEDA framework extracts the patch tokens from the initial blocks and trains separate CNNs on Clean+Attack samples (Adversarial Training) as shown in (1) and (2). A self-ensemble of these CNNs with the final ViT classifier goes through the distillation process with the new adversarial trained dataset to obtain a final distilled CNN model as seen in steps (3), (4), (5), and (6).

first m ($m < L$) blocks and combining their results with the final classification head. The SEViT ensemble can be formed by performing majority voting or randomly choosing only c out of the initial m intermediate classifiers. The SEViT defense mechanism also includes a detection mechanism, which aims to distinguish between the original image x and the perturbed image x', especially when the attack is successful ($f(x') \neq y$). The constraint on the distance metric between x and x' is that the L_∞ norm of their difference should be less than or equal to a predefined value epsilon. Although, we are not proposing any detection mechanism (to detect adversarial samples) in our enhanced solution as we can use the same detection approach as SEViT.

SEDA. Our two main hypotheses are: (1) Small CNN blocks would be more computationally efficient alternative to MLP blocks for each ViT block. (2) Defensive distillation when performed with adversarial training would make the model more robust against adversarial and model extraction attacks. Hence, we propose to enhance the existing SEViT by making 3 major modifications: We propose to substitute MLP blocks with CNN blocks instead to test for efficiency. Next, we perform adversarial training on the SEViT model instead of training it on just clean samples. Finally, we generate soft predictions to train a new distilled model, resulting in the SEDA framework (depicted in Fig. 1).

The proposed modifications are based on the fact that CNNs are more efficient than MLPs in learning spatial features from tokens through convolution operations at different levels of abstraction, which leads to improved generalization performance and reduced overfitting. Additionally, by training the model with adversarial examples, the model becomes better at handling perturbations during inference and generalizes better to new and unseen adversarial examples. The use of soft probabilities during distillation results in a smaller and more efficient model with faster inference time and reduced computational requirements. Moreover, the distilled model is more robust to adversarial attacks since the soft probabilities introduce uncertainty and variation in the output probabilities, making it harder for attackers to generate adversarial examples that can fool the model.

4   Experiments

Dataset. The experiments are performed on a chest x-ray dataset [14] that includes 7,000 images, and the classification task is binary, where each image is either Normal or TB. The dataset was split randomly, with 80% allocated for training, 10% for validation, and the remaining 10% for testing. This split ensures that the training, validation, and testing sets are mutually exclusive, and it enables the evaluation of the model's generalization ability to unseen images.

Attack Types. In our study, we utilize the Foolbox library [15] to create 3 different types of adversarial attacks, which are FGSM [2], PGD [16], and AutoPGD [17]. To generate attack samples with these algorithms, we use two values of perturbation budget $\epsilon = 0.03$ and $\epsilon = 0.003$ while keeping all other parameters at their default values.

MLPs Alternatives. To implement the Vision Transformer (ViT) model, we utilized the ViT-B/16 architecture pre-trained on ImageNet [5]. We used the fine-tuned ViT used in the original SEViT method. We experimented with several MLP alternatives with the aim to preserve clean accuracy and enhance robust accuracy while significantly reducing computational requirements. In order to create intermediate classifiers that take patch tokens as input, we trained 12 MLP alternatives such as CNN and ResNet variants for every block. In particular, we trained alternatives including a 2 convolution layer CNN, fine-tuned ResNet-50, transfer-learned ResNet-50, fine-tuned ResNet-50 with 2 additional convolution layers, and transfer-learned ResNet-50 with 2 additional convolution layers. Fine-tuning involves training the ResNet-50 model on our specific dataset, while transfer learning involves using the pre-trained weights from ImageNet and adapting the model to our specific task. Our experiments were conducted on a single Nvidia Quadro RTX 6000 GPU. All models are tested with the batch size of 30 and input size of $3 \times 224 \times 224$.

5   Results and Discussion

Accuracy and Robustness. Comparing the performances (Table 1 and Fig. 2) of our approach to the MLP block while ensembling consistently on 3 models (i.e., m = 3), as illustrated in Fig. 2, we conclude that the 2 convolution layer CNN alternative outperforms the original MLP block as it closely preserves the clean accuracy (with mean difference of $\leq 1\%$) and achieves higher accuracy against adversarial attack samples and that with significantly lower number of

Table 1. ViT vs SEViT vs SEViT-CNN (highlighted) in terms of Clean and Robust accuracy across the different numbers of intermediate ensembles. m = #Ensembles

Ensemble	m	(a). PRE-Adversarial Training							(b). POST-Adversarial Training						
		Clean	FGSM		PGD		AutoPGD		Clean	FGSM		PGD		AutoPGD	
			0.03	0.003	0.03	0.003	0.03	0.003		0.03	0.003	0.03	0.003	0.03	0.003
ViT [10]	-	96.38	55.65	91.59	32.62	92.17	23.77	92.46	-	-	-	-	-	-	-
MLP [4]	1	94.20	70.29	90.44	62.90	91.16	58.41	91.59	91.45	70.15	88.41	62.61	88.99	58.70	89.71
	2	**96.52**	83.91	95.22	80.15	95.36	78.41	**95.51**	95.51	84.06	93.48	81.01	93.62	80.87	93.91
	3	95.36	84.64	94.06	82.75	93.91	81.30	94.20	95.65	86.38	93.62	84.49	93.77	84.06	94.20
CNN (Ours)	1	93.04	70.87	89.86	64.78	90.58	60.44	90.73	95.51	71.30	92.03	63.77	92.75	58.84	92.75
	2	96.23	85.22	94.20	82.32	94.49	82.03	94.64	95.65	87.83	94.64	86.09	94.78	87.54	94.64
	3	96.23	**87.39**	**94.64**	**85.22**	**94.93**	84.93	94.93	**96.09**	**89.42**	**95.36**	**88.70**	**95.36**	**89.71**	**95.22**

parameters. This can be attributed to CNN's ability to learn spatial features from tokens through convolution operations at different levels of abstraction, leading to reduced overfitting and improved generalization performance. For example, when subjected to attack samples generated by FGSM (ϵ=0.03), the CNN alternative achieved 87.4% robust accuracy, which is higher than the 84.6% robust accuracy achieved by the 4-layered MLP block. When combined with distillation and adversarial training, the proposed SEDA framework is seen to achieve even higher clean and robust accuracy, exceeding an improvement of 6%. This increment in robust accuracy can also be noticed consistently on other attacks.

Computation. In contrast to SEViT, our proposed SEViT-CNN, and its distilled version, SEDA, offer significantly improved clean and robust accuracies for real-world deployment. SEDA is also 70 times more memory efficient than the original SEViT, as demonstrated by the computational and accuracy parameters shown in Table 4. The computational efficiency of a DL model can be evaluated based on various factors such as the FLOPs and accuracy. While the original MLP block in SEViT is accurate, it requires a significant amount of computational resources to train and deploy with having about 625M parameters [18] (Table 4). Based on these findings, we conclude that the CNN alternative is the best choice for ensembling ViT considering clean and robust accuracy, as well as computational efficiency.

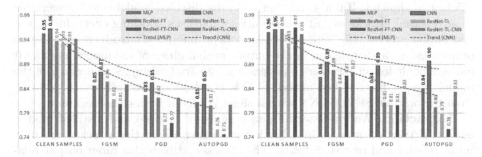

Fig. 2. Clean and adversarial performance of MLP alternatives Before adversarial training (left) vs After adversarial training (right). For each MLP alternative, the number of ensembles are m = 3 where the attack samples have $\epsilon = 0.03$

Table 2. Comparative performance of different ensembling models and their distilled versions (with #ensembles (m) = 3) against clean samples and attack samples. The reported results are the median of multiple runs.

| Model | (a). PRE-Adversarial Training | | | | | | | (b). POST-Adversarial Training | | | | | | |
| | Clean | FGSM | | PGD | | AutoPGD | | Clean | FGSM | | PGD | | AutoPGD | |
		0.03	0.003	0.03	0.003	0.03	0.003		0.03	0.003	0.03	0.003	0.03	0.003
ViT (m=0)	**96.38**	55.65	91.59	32.62	92.17	23.77	92.46	-	-	-	-	-	-	-
SEViT	95.36	84.64	94.06	82.75	93.91	81.30	94.20	95.65	86.38	93.62	84.49	93.77	84.06	94.20
SEViT-CNN (Ours)	96.23	87.39	**94.64**	85.22	94.93	84.93	94.93	**96.09**	89.42	95.36	88.70	95.36	89.71	95.22
Distilled (ViT)	94.71	84.64	87.10	86.67	93.33	**90.87**	94.20	-	-	-	-	-	-	-
Distilled (SEViT)	92.71	84.78	88.26	86.52	93.33	88.70	93.91	93.86	87.39	**97.68**	87.83	**97.97**	90.87	**96.67**
SEDA (Ours)	94.43	**88.84**	94.20	**88.70**	**95.65**	90.29	**95.94**	94.86	**89.86**	96.38	**90.87**	96.09	**92.46**	96.09

Table 3. Comparison of extracted model when model extraction is performed on original model V/s when extraction is performed on the distilled versions (with m = 3). From defender's view, *Lower* clean/Adv. accuracy is better. Note, SEDA (highlighted) = SEViT-CNN + Distillation + Post Adversarial Training.

| Extraction On | (a). PRE-Adversarial Training | | | | | | | (b). POST-Adversarial Training | | | | | | |
| | Clean | FGSM | | PGD | | AutoPGD | | Clean | FGSM | | PGD | | AutoPGD | |
		0.03	0.003	0.03	0.003	0.03	0.003		0.03	0.003	0.03	0.003	0.03	0.003
ViT (m=0)	84.57	82.75	85.65	82.61	85.65	83.77	85.94	-	-	-	-	-	-	-
SEViT	84.43	83.62	85.51	83.04	85.51	83.62	85.51	83.57	82.32	84.93	82.90	84.93	82.46	85.07
SEViT-CNN (Ours)	84.57	81.30	84.64	81.74	84.78	82.32	85.36	**81.43**	80.43	82.61	80.29	82.75	80.14	82.90
Distilled (ViT)	85.14	83.04	85.65	82.32	85.80	82.32	86.23	-	-	-	-	-	-	-
Distilled (SEViT)	83.57	78.99	**83.77**	79.13	84.60	79.75	84.64	84.57	82.90	85.36	82.32	85.36	83.19	85.80
SEDA (Ours)	**83.71**	**79.86**	84.20	79.71	**84.20**	**79.71**	84.64	81.86	82.03	83.04	82.17	83.04	82.03	83.48

5.1 Defensive Distillation vs Extraction Attack

Distillation. Table 2 indicates that the distilled model outperformed the original model in terms of robust accuracy despite having a smaller architecture (of 5 convolution layer CNN). However, there was a slight decline of around 2% in the clean accuracy of the distilled model, which could be viewed as a reasonable trade-off between clean and robust accuracy. Distillation using soft probabilities leads to a compact and smaller architecture, which results in faster inference time and lower computational requirements, making the model more efficient as shown in Table 4. Furthermore, the distilled model is less susceptible to adversarial attacks as soft probabilities impart a smoothing effect during the distillation process [19]. This is because soft probabilities introduce some degree of uncertainty and variation in the output probabilities of the model, making it more difficult for an attacker to generate adversarial examples that can fool the model. Hence, the distilled model, SEDA, is robust against adversarial attacks than the original SEViT-CNN with a slight trade-off with clean accuracy.

Extraction. We evaluate the performance of original and distilled models against extraction attacks in a black-box setting where the attacker can only input queries and receive outputs. The attacker utilizes this input-output relationship to create a replica/attack/extracted model (a 3 convolution layer CNN) of the original model. The aim is to compare the extracted model's performance in two scenarios: one where the original model is deployed, and the other where

Table 4. Extensive computational analysis of the SEDA and its comparison with the alternative models in terms of computational parameters. *Distilled* models are distilled with #ensembles (m) = 3. The '+' refers to *in addition* to ViT parameters.

Model	#Params	FLOPs	Weight	Inference Time	Throughput	Clean Accuracy	Adv. Acc. (ϵ=0.03)			
							FGSM	PGD	AutoPGD	Mean
ViT (m=0)	85.64M	16.86G	327.37MB	7.06	277.25	**96.38**	55.65	32.62	23.77	37.35
SEViT (m=3)	+1875.66M	+1875.63M	+6.99GB	12.87	13125.57	**95.36**	84.64	82.75	81.30	82.90
SEViT-CNN (m=3)	+3.09M	+531.27M	+11.82MB	1.20	124761.51	**96.23**	87.39	85.22	84.93	85.85
Distilled (ViT)	27.79M	994.41M	106.00MB	0.96	3956.04	94.71	84.64	86.67	90.87	87.39
Distilled (SEViT)	27.79M	994.41M	106.00MB	0.97	3952.44	92.71	84.78	86.52	88.70	86.67
SEDA (Ours)	**27.79M**	**994.41M**	**106.00MB**	0.96	3962.13	**94.86**	**89.86**	**90.87**	**92.46**	**91.06**

the distilled variant is deployed. This comparison allows us to determine that deploying which model would be more resilient to model extraction attacks. Table 3 shows that the attacker's clean accuracy (post-adversarial training) in reproducing the distilled SEViT-CNN, i.e., SEDA (81.86%) is lower compared to that of the original SEViT-CNN (84.57%) and SEViT (84.43%). The models extracted by the attacker, exhibit lower level of accuracy and robustness than the original model, with SEViT-CNN and SEDA being among the lowest, implying that the attacker is less likely to recover the exact parameters of these models due to the smoothed probabilities. Thus, we can conclude that deploying the distilled model is a more secure option against extraction attacks than the original SEViT or ViT model.

5.2 Adversarial Training

Accuracy and Robustness. In the case of pre-adversarial training, both models (Original and Distilled) have high accuracy on clean samples, but their performance on adversarial examples is increased even higher following post-adversarial training, by at least +2 to +5% (Table 1 and Table 2). Additionally, the models' vulnerability to model extraction attacks is high in this pre-adversarial training case, with the extracted models' accuracy on clean and adversarial examples being considerably high. However, following post-adversarial training, models extracted on original and their distilled versions showed a considerable decrement (of about -3%) in their clean and robust accuracies (Table 3). These results demonstrate the improved robustness of the models via adversarial training with improved generalization and their increased resistance against attacks and model extraction.

6   Conclusion and Future Work

In this work, we have proposed an improved architecture, Self-Ensembling ViT with defensive Distillation and Adversarial training (SEDA), to overcome computational bottlenecks and improve robustness by making 3 major modifications: we propose to substitute MLP blocks with CNN instead to test for efficiency.

Next, we perform adversarial training on the SEViT-CNN model instead of training it on just clean samples. And last, we generate soft predictions to train a new distilled model. This study is the first to combine these techniques with ViT ensembling, providing a SOTA defense for ViTs against adversarial attacks and model extraction attacks. We prove the effectiveness of our novel enhanced approach: SEDA, using an extensive set of experiments on publicly available Tuberculosis x-ray dataset.

In the future, we aim to further enhance the proposed SEDA model using (i) defensive approaches such as differential privacy (ii) exploring the trade-off between adversarial robustness and model accuracy when performing defensive distillation (iii) exploring a diverse set of alternatives for each block rather than just having CNN for every block. This would increase the diversity among the ensemble models and further improve overall performance.

References

1. Çallı, E., Sogancioglu, E., van Ginneken, B., G van Leeuwen, K., Murphy, K.: Deep learning for chest x-ray analysis: a survey. Med. Image Anal. **72**, 102125 (2021)
2. Goodfellow, I., Shlens, J., Szegedy, C.: Explaining and harnessing adversarial examples. In: International Conference on Learning Representations (2015)
3. Huang, Q.-X., Yap, W.L., Chiu, M.-Y., Sun, H.-M.. Privacy-preserving deep learning with learnable image encryption on medical images. IEEE Access **10**, 66345–66355 (2022)
4. Almalik, K., Yaqub, M., Nandakumar, K.: Self-ensembling vision transformer (sevit) for robust medical image classification. In International Conference on Medical Image Computing and Computer-Assisted Intervention, pp. 376–386. Springer, Cham (2022). https://doi.org/10.1007/978-3-031-16437-8_36
5. Imam, R., Huzaifa, M., El-Amine Azz, M.: On enhancing the robustness of vision transformers: Defensive diffusion. arXiv preprint arXiv:2301.13188 (2023)
6. Kaviani, S., Han, K.J., Sohn, I.: Adversarial attacks and defenses on ai in medical imaging informatics: a survey. Expert Syst. Appl., 116815 (2022)
7. Carlini, N., et al.: Extracting training data from diffusion models. arXiv preprint arXiv:2301.13188 (2023)
8. Rasool, R.U., Ahmad, H.F., Rafique, W., Qayyum, A., Qadir, J.: Security and privacy of internet of medical things: a contemporary review in the age of surveillance, botnets, and adversarial ml. J. Network Comput. Appl., 103332 (2022)
9. Papernot, N., McDaniel, P., Wu, X., Jha, S., Swami, A.: Distillation as a defense to adversarial perturbations against deep neural networks. In: 2016 IEEE Symposium on Security and Privacy (SP), pp. 582–597. IEEE (2016)
10. Vaswani, A., et al.: Attention is all you need. Advances in Neural Information Processing Systems, 30 (2017)
11. Naseer, M., Ranasinghe, K., Khan, S., Khan, F.S., Porikli, F.: On improving adversarial transferability of vision transformers. In: The Tenth International Conference on Learning Representations (2022)
12. Malik, H.S., Kunhimon, S., Naseer, M., Khan, S., Khan, F.S.: Adversarial pixel restoration as a pretext task for transferable perturbations. In 33rd British Machine Vision Conference 2022, BMVC 2022, London, UK, November 21–24, 2022. BMVA Press (2022)

13. Wu, B., Gu, J., Li, Z., Cai, D., He, X., Liu, W.: Towards efficient adversarial training on vision transformers. In Computer Vision-ECCV 2022: 17th European Conference, Tel Aviv, Israel, October 23–27, 2022, Proceedings, Part XIII, pp. 307–325. Springer, Cham (2022). https://doi.org/10.1007/978-3-031-19778-9_18
14. Rahman, T., et al.: Reliable tuberculosis detection using chest x-ray with deep learning, segmentation and visualization. IEEE Access **8**, 191586–191601 (2020)
15. Rauber, J., Zimmermann, R., Bethge, M., Brendel, W.: Foolbox native: Fast adversarial attacks to benchmark the robustness of ml models in pytorch, tensorflow, and jax. Journal of Open Source Software **5**(53), 2607 (2020)
16. Madry, A., Makelov, A., Schmidt, L., Tsipras, D., Vladu, A.: Towards deep learning models resistant to adversarial attacks. In: International Conference on Learning Representations (2018)
17. Francesco Croce and Matthias Hein. Reliable evaluation of adversarial robustness with an ensemble of diverse parameter-free attacks. In International Conference on Machine Learning, pages 2206–2216. PMLR, 2020
18. Xin Zhou, Zhepei Wang, Xiangyong Wen, Jiangchao Zhu, Chao Xu, and Fei Gao. Decentralized spatial-temporal trajectory planning for multicopter swarms. arXiv preprint arXiv:2106.12481, 2021
19. Nicholas Carlini and David Wagner. Towards evaluating the robustness of neural networks. In 2017 IEEE Symposium on Security and Privacy (sp), pages 39–57. IEEE, 2017

A Continual Learning Approach for Cross-Domain White Blood Cell Classification

Ario Sadafi[1,2], Raheleh Salehi[1,3], Armin Gruber[1,4],
Sayedali Shetab Boushehri[1,5], Pascal Giehr[3], Nassir Navab[2,6],
and Carsten Marr[1(✉)]

[1] Institute of AI for Health, Helmholtz Munich – German Research Center for
Environmental Health, Neuherberg, Germany
carsten.marr@helmholtz-munich.de
[2] Computer Aided Medical Procedures (CAMP), Technical University of Munich,
Munich, Germany
[3] Department of Chemistry, Institute for Chemical Epigenetics Munich (ICEM),
Ludwig-Maximilians University Munich, Munich, Germany
[4] Laboratory of Leukemia Diagnostics, Department of Medicine III, University
Hospital, Ludwig-Maximilians University Munich, Munich, Germany
[5] Data and Analytics, Pharmaceutical Research and Early Development, Roche
Innovation Center Munich (RICM), Penzberg, Germany
[6] Computer Aided Medical Procedures, Johns Hopkins University, Baltimore, USA

Abstract. Accurate classification of white blood cells in peripheral
blood is essential for diagnosing hematological diseases. Due to con-
stantly evolving clinical settings, data sources, and disease classifica-
tions, it is necessary to update machine learning classification models
regularly for practical real-world use. Such models significantly benefit
from sequentially learning from incoming data streams without forgetting previously acquired knowledge. However, models can suffer from
catastrophic forgetting, causing a drop in performance on previous tasks
when fine-tuned on new data. Here, we propose a rehearsal-based continual learning approach for class incremental and domain incremental
scenarios in white blood cell classification. To choose representative samples from previous tasks, we employ exemplar set selection based on the
model's predictions. This involves selecting the most confident samples
and the most challenging samples identified through uncertainty estimation of the model. We thoroughly evaluated our proposed approach on
three white blood cell classification datasets that differ in color, resolution, and class composition, including scenarios where new domains or
new classes are introduced to the model with every task. We also test
a long class incremental experiment with both new domains and new
classes. Our results demonstrate that our approach outperforms established baselines in continual learning, including existing iCaRL and EWC
methods for classifying white blood cells in cross-domain environments.

A. Sadafi and R. Salehi—Equal contribution.

Keywords: Continual learning · Single blood cell classification · Epistemic uncertainty estimation

1 Introduction

Microscopic examination of blood cells is an essential step in laboratories for diagnosing haematologic malignancies and can be aided by computational diagnostic tools. So far, the developed classification methods have performed quite well. For instance, Matek et al. [19] proposed an expert-level tool for recognizing blast cells in blood smears of acute myeloid leukemia patients that also works for accurately classifying cell morphology in bone marrow smears [17]. Several contributions are addressing the problem also at disease diagnosis level [11,23]. Boldu et al. [2] have proposed an approach for diagnosing acute leukemia from blast cells in blood smear images. Sidhom et al. [25] and Eckardt et al. [7,8] are proposing methods based on deep neural networks to detect acute myeloid leukemia, acute promyelocytic leukemia and NPM1 mutation from blood smears of patients. All of these methods are highly dependent on the data source. Laboratories with different procedures and microscope settings produce slightly different images, and adapting models [22] to new streams of data arriving on a regular basis is a never-ending task. Although some domain adaptation methods have been proposed to increase the cross-domain reusability, for instance, the cross-domain feature extraction by Salehi et al. [24], regular finetuning of models on new data in each center is still a necessity.

When a model is trained on new data, catastrophic forgetting can occur in neural network models leading to loss of information it had learned from the previous data [12]. Continual learning techniques are designed to tackle catastrophic forgetting and might become a necessity in clinical applications [14]. They can be divided into three main categories: (i) Task-specific components where different sub-networks are designed for each task. (ii) Regularized optimization methods such as elastic weight consolidation (EWC) [13] that control the update of weights to minimize forgetting. (iii) Rehearsal based approaches such as iCaRL [21] where a small memory is used to keep examples from previous tasks to augment new datasets.

In medical image analysis, however, these continual learning techniques are less frequent. Li et al. [16] have developed a method based on continual learning in task incremental scenarios for incremental diagnosis systems. For chest X-ray classification, continual learning has been applied to the domain adaptation problem by Lenga et al. [15] or on domain incremental learning by Srivastava et al. [26]. In another work, Derakhshani et al. [6] benchmarked different continual learning methods on MNIST-like collections of biomedical images. Their study highlights the potential of rehearsal-based approaches in disease classification.

In this paper, we are proposing an uncertainty-aware continual learning (UACL) method, a powerful approach for both class incremental and domain incremental scenarios. We test our rehearsal-based method on three real-world datasets for white blood cell classification, demonstrating its potential. We use a distillation loss to preserve the previously learned information and provide a sampling strategy based on both representativeness of examples for each class and the uncertainty of the model. In order to foster reproducible research, we are providing our source code at https://github.com/marrlab/UACL

2   Methodology

2.1   Problem Formulation

Let $\mathcal{D}_t = \{x_i, y_i\}$ denote the training set at step t with images x_i and their corresponding labels y_i. The classifier $f_t(.)$ is initialized with model parameters θ_{t-1} learned during the previous stage. To facilitate the new task (such as classifying a new class), additional parameters are introduced at every stage, which are randomly initialized. Our incremental learning method updates the model parameters θ_t for the new task based on the training set $\mathcal{D}_t$ and an exemplar set X_t. The exemplar set contains selected sample images from the previous steps.

In order to mitigate catastrophic forgetting, in UACL we suggest: (i) At every stage, the training is augmented with the exemplar set X_t consisting of the sampled images from previous tasks. (ii) A distillation loss is added to preserve the information learned in the previous stage.

2.2   Classifier Model

Our method is architecture agnostic, and any off-the-shelf classifier can be used. We decided to follow Matek et al. [19] and use a ResNeXt-50 [28] as the single cell classifier. ResNeXt is an extension to the ResNet [10] architecture introducing the cardinality concept defined as the number of parallel paths within a building block. Each path contains a set of aggregated transformations allowing ResNeXt to capture fine-grained and diverse features.

2.3   Training

At every training step, the new stream of data $\mathcal{D}_t$ is combined with the examples sampled from the previous steps X_t. The training is performed by minimizing a total loss term $\mathcal{L}$ consisting of classification and distillation loss as follows:

$$\mathcal{L}(\theta_t) = \mathcal{L}_{\mathrm{cls}}(\theta_t) + \gamma \mathcal{L}_{\mathrm{dist}}(\theta_t) \tag{1}$$

where γ is a coefficient controlling the effect of distillation on the total loss. For classification loss, we are using categorical cross-entropy loss defined as

$$\mathcal{L}_{\mathrm{cls}}(\theta_t) = -\sum_{j \in C_t} y_i^{(j)} \log(f(x_i; \theta_t)^{(j)}) \qquad \forall (x_i, y_i) \in \mathcal{D}_t \cup X_t \tag{2}$$

where C_t is the set of all classes that the model has been exposed to since the beginning of the experiment. The distillation loss is defined between the current state of the model with θ_t parameters and the previous state with learned θ_{t-1} parameters. Using binary cross-entropy loss, we try to minimize the disagreement between the current model and the model before training:

$$\mathcal{L}_{\text{dist}}(\theta_t) = \sum_{j \in C_{t-1}} \mathcal{S}(f(x_i; \theta_{t-1})).\log(\mathcal{S}(f(x_i; \theta)))$$

$$+ (1 - \mathcal{S}(f(x_i; \theta_{t-1}))).\log(1 - \mathcal{S}(f(x_i; \theta)))$$

$$\forall (x_i, y_i) \in \mathcal{D}_t \cup X_t \quad (3)$$

where $\mathcal{S}(x) = (1 + \exp(x))^{-1}$ is the sigmoid function.

2.4   Exemplar Set Selection

Selection of the exemplar set is crucial for the performance of our uncertainty aware continual learning method. We assume a memory budget equal to K images, so for every class $m = K/|C_t|$ images can be selected. We use two criteria for the selection of the images:

(i) $m/2$ of the images are selected according to their distance from the class mean in the feature space such that the images that are best approximating the class mean are preserved. If $\psi(x; \theta_t)$ returns the feature vector of the model at the last convolutional layer of the model for input x, each image p_i is selected iteratively using the following relation

$$p_i = \underset{x \in \mathcal{P}}{\operatorname{argmin}} \left\| \frac{1}{n} \sum_{x_i \in \mathcal{P}} \psi(x_i) - \frac{1}{k}[\psi(x) + \sum_{j=1}^{k-1} \psi(p_j)] \right\| \quad (4)$$

where $\mathcal{P}$ is the set of all images belonging to the class and $1 < k < m/2$ is the total number of selected images so far.

(ii) We use epistemic uncertainty estimation [20] to select the remaining $m/2$ images. We estimate the uncertainty of the model on every image by measuring its predictive variance across T inferences, using a dropout layer introduced between the second and third layer of the architecture. This dropout serves as a Bayesian approximation [9]. For every image, the model predicts $|C_t|$ probabilities resulting into a matrix $R^{T \times |C_t|}$. Thus for every image x_i, the uncertainty is calculated as

$$U_i = \sum_{j=1}^{|C|} \sqrt{\frac{\sum_{k=1}^{T}(R_{kj}^{x_i} - \mu_j^{x_i})^2}{T}} \qquad \forall x_i \in \mathcal{P} \quad (5)$$

where $\mu_j^{x_i} = \frac{1}{T}\sum_i^T R_i^{x_i}$ is the vector with mean values of R for each class j. After this step, images are sorted based on the uncertainty scores, and the $m/2$ images with the highest uncertainty are selected.

2.5   Exemplar Set Reduction

Since the total amount of memory is limited, in case new classes or domains are introduced, the method needs to shrink the currently preserved samples in order to accommodate the new classes. Since both lists of class mean contributors and uncertain images are sorted, this reduction is made by discarding the least important samples from the end of the list.

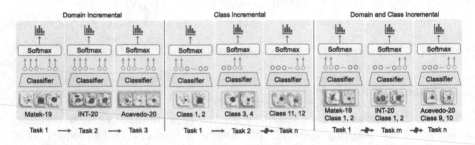

Fig. 1. Overview of the three different experiments we tested our proposed uncertainty aware continual learning (UACL) method: In domain incremental, in each task, the classifier is exposed to all classes of the new dataset. In class incremental, the scenario is repeated for each dataset, gradually introducing the classes to the model. Lastly, the performance of the approach is tested on a combination of domain and class incremental, where classes are gradually introduced to the model and continued with all datasets.

3   Evaluation

3.1   Datasets and Continual Learning Experiments

Three single white blood cell datasets from different sources are used for our continual learning experiments. Matek-19 [3,18] is a single white blood cell collection from 100 acute myeloid leukemia patients and 100 non-leukemic patients from Munich University Hospital between 2014 and 2017, and publicly available INT-20 is an in-house dataset with over 23,000 annotated white blood cells. Acevedo-20 is another public dataset [1] with over 14,000 images, acquired at the core laboratory at the Hospital Clinic of Barcelona. For further details about the datasets and train test split, refer to Table 1.

We define three different continual learning scenarios to evaluate our method:

- **Domain incremental**: With three datasets, three different tasks are defined for training, where the model is exposed only to one of the datasets at a time and tested on the rest of the unseen datasets. The goal is to preserve the class representations while the domain changes.

- **Class incremental**: In this scenario, four tasks are defined, each containing two to four new classes for the model to learn (see Table 1). This scenario repeats for all three datasets.

Table 1. Statistics of the three datasets used for experiments.

Datasets	Number of classes	Train set images	Test set images	Classes per task	Image size (pixels)
Matek-19	13	12,042	3,011	3,3,3,4	$400 \times 400_{(29 \times 29\,\mu m)}$
INT-20	13	18,466	4,617	3,3,3,4	$288 \times 288_{(25 \times 25\,\mu m)}$
Acevedo-20	10	11,634	2,909	3,2,2,3	$360 \times 363_{(36 \times 36.3\,\mu m)}$

- Domain and class incremental: This is the longest continual learning scenario we could define with the datasets. Starting with Matek-19, the model is exposed to four tasks in a class incremental fashion. After that, the experiment continues with another 4 tasks from INT-20 followed by 4 more tasks from Acevedo-20 resulting in 12 tasks across the three datasets.

Figure 1 provides an overview of the scenarios, and Table 1 shows statistics about the defined tasks.

Table 2. Average accuracy and average forgetting is reported for all methods and all experiments. Finetuning and EWC perform on par in almost all experiments, while our approach consistently outperforms the other baselines. Abbreviations in experiment column are D: Domain incremental, C_M: Class incremental Matek-19, C_I: Class incremental INT-20, C_A: Class incremental Acevedo-20, DC: Domain and class incremental. An upper bound (UB) is also provided where models have access to all data.

	UACL		Finetuning		EWC		iCaRL		UB
	Acc ↑	For ↓	Acc ↑	For ↓	Acc ↑	For ↓	Acc ↑	For ↓	Acc↑
D	**0.82**±0.04	**0.01**±0.02	0.57±0.28	0.47±0.48	0.58±0.28	0.48±0.48	0.80±0.09	0.06±0.03	0.96±0.01
C_M	**0.74**±0.17	**0.06**±0.10	0.47±0.23	0.28±0.39	0.46±0.20	0.24±0.26	0.62±0.18	0.08±0.12	0.89±0.05
C_I	**0.85**±0.11	**0.08**±0.03	0.51±0.29	0.33±0.47	0.65±0.28	0.33±0.13	0.81±0.15	0.11±0.08	0.97±0.02
C_A	**0.83**±0.11	0.05±0.04	0.49±0.30	0.33±0.47	0.61±0.26	0.16±0.05	0.81±0.13	**0.04**±0.01	0.97±0.02
DC	**0.80**±0.06	**0.01**±0.05	0.22±0.23	0.08±0.24	0.23±0.23	0.07±0.26	0.52±0.16	0.06±0.17	0.75±0.16

Baselines. For comparison, we are using three different baselines: (i) Simple fine-tuning of the model as a naive baseline. (ii) Elastic weight consolidation (EWC) [13] method imposes penalties on change of certain weights of the network that are important for the previously learned tasks while allowing other less important weights to be updated during training to accommodate the learning of the new task. (iii) Incremental classifier and representation learning (iCaRL) [21], a method combining feature extraction, nearest neighbor classification, and exemplar storage for rehearsing previous tasks.

Implementation Details. We divided the datasets into stratified 75% training set and 25% test set. Models are trained for 30 epochs with stochastic gradient descent, a learning rate of $5e-4$, and a Nestrov momentum [27] of 0.9. Models are initialized with ImageNet [5] weights for the training of the first task. For EWC

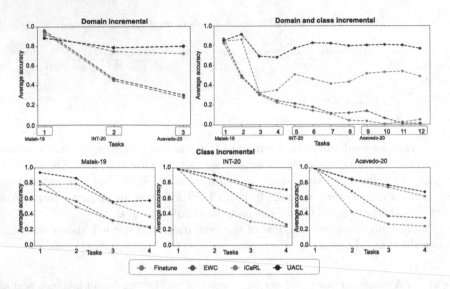

Fig. 2. We compare our proposed method with three different approaches as baselines for all three experiments regarding average accuracy. Our UACL method outperforms all other methods and is on par with iCaRL in domain incremental and class incremental experiments.

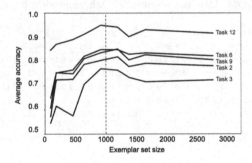

Fig. 3. Average accuracy over the memory of the total exemplar sets (K) in domain class incremental experiment. We conducted our experiments at K = 1000.

experiments, the learning rate is set to $3e-4$ and lambda to 0.9 to weight EWC penalty loss. EWC penalty importance is set to 3000. For iCaRL experiments, we decided to have an exemplar size $K = 1000$ and $\gamma = 1$ for 3-dataset experiments (i.e., domain incremental and class-domain incremental experiments), while $K = 125$ and $\gamma = 5$ for class incremental experiments on single datasets. We used 10 forward passes for MC dropout sampling. We use average accuracy [4] and average forgetting [4] for the evaluation of all experiments.

3.2   Results and Discussion

We have conducted five experiments at class incremental and domain incremental learning (see Table 2). In addition to the defined baselines (Finetuning, EWC, and iCaRL), we are also comparing our UACL method to the hypothetical scenario where the model has access to complete datasets from previous tasks, as an upper bound (UB) to the problem. Our UACL method outperforms all other baselines in average accuracy (Acc). It outperforms all other baselines also in average forgetting (For), apart from the class incremental Acevedo-20 experiment, where iCaRL is slightly better (0.04 vs. 0.05, see Table 2).

Our UACL method is specifically good for long-term learning. According to Fig. 2, UACL outperforms all other methods in the domain class incremental experiments due to its epistemic uncertainty-based sampling that correctly samples the most challenging examples from every class. Interestingly, we are outperforming the upper bound in the domain class incremental experiment, which can be due to the unbalanced data (see Table 2). With the UACL sampling strategy, classes are equally sampled, and uncertain cases in underrepresented classes are very informative to preserve the latent structure of these classes.

An important hyperparameter in the experiments is the exemplar set size K. We conducted an experiment studying the effect of K on the average accuracy metric. Figure 3 demonstrates the average accuracy for different values of K for different tasks in the domain and class incremental experiment. As expected, small values of the exemplar set size are unable to capture the variability present in the classes. However, as the value of K increases, the method reaches higher average accuracy and a plateau. Specifically, the experiment indicates that an exemplar set size of around $K = 1000$ achieves the optimal value for capturing class variability and achieving high accuracy while using less memory.

Additionally, we tried to qualitatively study the difference in the UACL sampling method we are suggesting. Figure 4 shows the UMAP of the features on the penultimate layer of the ResNext model. We have zoomed into some of the class distributions and show the difference between our sampling approach (orange) and iCaRL's class mean preservation approach (blue). Our uncertainty-based sampling leads to a better distribution of the samples within a class. We believe scattered points over the whole class distribution and closer points to class distribution borders are essentials for long-term continual learning.

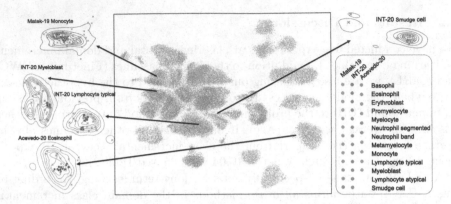

Fig. 4. Epistemic uncertainty-based sampling better preserves the latent distribution of the classes. This is demonstrated with the UMAP embedding of the training set with a total of 52,629 cells. Each class in each dataset has a different color. For some classes, the kernel distribution estimation plot and the selected examples by epistemic UCAL sampling (orange) and iCaRL-based sampling (blue) are displayed. (Color figure online)

4   Conclusion

Accurate classification of white blood cells is crucial for diagnosing hematological diseases, and continual learning is necessary to update models regularly for practical, real-world use with data coming from different domains or new classes. To address this issue, we proposed a rehearsal-based continual learning approach that employs exemplar set selection based on the model's uncertainty to choose representative samples from previous tasks. We defined three continual learning scenarios for comparison. The results indicate that our proposed method outperforms established baselines in continual learning, including iCaRL and EWC methods, for classifying white blood cells. Collecting larger and better curated datasets to perform chronological incremental learning where data is collected at different time points in a center, performing longer experiments to investigate the scalability of our proposed method, and investigation of other uncertainty estimation methods are some of the exciting directions for possible future works.

Acknowledgments. C.M. has received funding from the European Research Council (ERC) under the European Union's Horizon 2020 research and innovation programme (Grant agreement No. 866411). R.S. and P.G. are funded by DFG, SFB1309 - project number 325871075.

References

1. Acevedo, A., Merino, A., Alférez, S., Molina, Á., Boldú, L., Rodellar, J.: A dataset of microscopic peripheral blood cell images for development of automatic recognition systems. Data Brief **30**, 105474 (2020)

2. Boldú, L., Merino, A., Alférez, S., Molina, A., Acevedo, A., Rodellar, J.: Automatic recognition of different types of acute leukaemia in peripheral blood by image analysis. J. Clin. Pathol. **72**(11), 755–761 (2019)

3. Clark, K., et al.: The Cancer Imaging Archive (TCIA): maintaining and operating a public information repository. J. Digit. Imaging **26**, 1045–1057 (2013)

4. De Lange, M., et al.: A continual learning survey: defying forgetting in classification tasks. IEEE Trans. Pattern Anal. Mach. Intell. **44**(7), 3366–3385 (2021)

5. Deng, J., Dong, W., Socher, R., Li, L.J., Li, K., Fei-Fei, L.: ImageNet: a large-scale hierarchical image database. In: 2009 IEEE Conference on Computer Vision and Pattern Recognition, pp. 248–255. IEEE (2009)

6. Derakhshani, M.M., et al.: LifeLonger: a benchmark for continual disease classification. In: Wang, L., Dou, Q., Fletcher, P.T., Speidel, S., Li, S. (eds.) International Conference on Medical Image Computing and Computer-Assisted Intervention. LNCS, vol. 13432, pp. 314–324. Springer, Cham (2022). https://doi.org/10.1007/978-3-031-16434-7_31

7. Eckardt, J.N., et al.: Deep learning detects acute myeloid leukemia and predicts NPM1 mutation status from bone marrow smears. Leukemia **36**(1), 111–118 (2022)

8. Eckardt, J.N., et al.: Deep learning identifies Acute Promyelocytic Leukemia in bone marrow smears. BMC Cancer **22**(1), 1–11 (2022)

9. Gal, Y., Ghahramani, Z.: Dropout as a Bayesian approximation: representing model uncertainty in deep learning. In: International Conference on Machine Learning, pp. 1050–1059. PMLR (2016)

10. He, K., Zhang, X., Ren, S., Sun, J.: Deep residual learning for image recognition. In: Proceedings of the IEEE Conference on Computer Vision and Pattern Recognition, pp. 770–778 (2016)

11. Hehr, M., et al.: Explainable AI identifies diagnostic cells of genetic AML subtypes. PLOS Digital Health **2**(3), e0000187 (2023)

12. Kemker, R., McClure, M., Abitino, A., Hayes, T., Kanan, C.: Measuring catastrophic forgetting in neural networks. In: Proceedings of the AAAI Conference on Artificial Intelligence, vol. 32 (2018)

13. Kirkpatrick, J., et al.: Overcoming catastrophic forgetting in neural networks. Proc. Natl. Acad. Sci. **114**(13), 3521–3526 (2017)

14. Lee, C.S., Lee, A.Y.: Clinical applications of continual learning machine learning. Lancet Digital Health **2**(6), e279–e281 (2020)

15. Lenga, M., Schulz, H., Saalbach, A.: Continual learning for domain adaptation in chest x-ray classification. In: Medical Imaging with Deep Learning, pp. 413–423. PMLR (2020)

16. Li, Z., Zhong, C., Wang, R., Zheng, W.-S.: Continual learning of new diseases with dual distillation and ensemble strategy. In: Martel, A.L., et al. (eds.) MICCAI 2020. LNCS, vol. 12261, pp. 169–178. Springer, Cham (2020). https://doi.org/10.1007/978-3-030-59710-8_17

17. Matek, C., Krappe, S., Münzenmayer, C., Haferlach, T., Marr, C.: Highly accurate differentiation of bone marrow cell morphologies using deep neural networks on a large image data set. Blood J. Am. Soc. Hematol. **138**(20), 1917–1927 (2021)

18. Matek, C., Schwarz, S., Marr, C., Spiekermann, K.: A single-cell morphological dataset of leukocytes from AML patients and non-malignant controls (AML-Cytomorphology_LMU). The Cancer Imaging Archive (TCIA) [Internet] (2019)

19. Matek, C., Schwarz, S., Spiekermann, K., Marr, C.: Human-level recognition of blast cells in acute myeloid leukaemia with convolutional neural networks. Nat. Mach. Intell. **1**(11), 538–544 (2019)

20. Mukhoti, J., Kirsch, A., van Amersfoort, J., Torr, P.H., Gal, Y.: Deterministic neural networks with appropriate inductive biases capture epistemic and aleatoric uncertainty. arXiv preprint arXiv:2102.11582 (2021)
21. Rebuffi, S.A., Kolesnikov, A., Sperl, G., Lampert, C.H.: iCaRL: incremental classifier and representation learning. In: Proceedings of the IEEE Conference on Computer Vision and Pattern Recognition, pp. 2001–2010 (2017)
22. Sadafi, A., et al.: Multiclass deep active learning for detecting red blood cell subtypes in brightfield microscopy. In: Shen, D., et al. (eds.) MICCAI 2019. LNCS, vol. 11764, pp. 685–693. Springer, Cham (2019). https://doi.org/10.1007/978-3-030-32239-7_76
23. Sadafi, A., et al.: Attention based multiple instance learning for classification of blood cell disorders. In: Martel, A.L., et al. (eds.) MICCAI 2020. LNCS, vol. 12265, pp. 246–256. Springer, Cham (2020). https://doi.org/10.1007/978-3-030-59722-1_24
24. Salehi, R., et al.: Unsupervised cross-domain feature extraction for single blood cell image classification. In: Wang, L., Dou, Q., Fletcher, P.T., Speidel, S., Li, S. (eds.) International Conference on Medical Image Computing and Computer-Assisted Intervention. LNCS, vol. 13433, pp. 739–748. Springer, Cham (2022). https://doi.org/10.1007/978-3-031-16437-8_71
25. Sidhom, J.W., et al.: Deep learning for diagnosis of acute promyelocytic leukemia via recognition of genomically imprinted morphologic features. NPJ Precis. Oncol. 5(1), 38 (2021)
26. Srivastava, S., Yaqub, M., Nandakumar, K., Ge, Z., Mahapatra, D.: Continual domain incremental learning for chest x-ray classification in low-resource clinical settings. In: Albarqouni, S., et al. (eds.) Domain Adaptation and Representation Transfer (DART), MICCAI. LNCS, vol. 12968, pp. 226–238. Springer, Cham (2021). https://doi.org/10.1007/978-3-030-87722-4_21
27. Sutskever, I., Martens, J., Dahl, G., Hinton, G.: On the importance of initialization and momentum in deep learning. In: International Conference on Machine Learning, pp. 1139–1147. PMLR (2013)
28. Xie, S., Girshick, R., Dollár, P., Tu, Z., He, K.: Aggregated residual transformations for deep neural networks. In: Proceedings of the IEEE Conference on Computer Vision and Pattern Recognition, pp. 1492–1500 (2017)

Metadata Improves Segmentation Through Multitasking Elicitation

Iaroslav Plutenko[1,2]($\boxtimes$), Mikhail Papkov[3], Kaupo Palo[4], Leopold Parts[3,5], and Dmytro Fishman[3]

[1] Ukrainian Catholic University, Lviv, Ukraine
plutenko@ipk-gatersleben.de
[2] Leibniz Institute of Plant Genetics and Crop Plant Research (IPK), Gatersleben, Germany
[3] Institute of Computer Science, University of Tartu, Tartu, Estonia
dmytro.fishman@ut.ee
[4] Revvity, Inc., Tallinn, Estonia
[5] Wellcome Sanger Institute, Hinxton, UK

Abstract. Metainformation is a common companion to biomedical images. However, this potentially powerful additional source of signal from image acquisition has had limited use in deep learning methods, for semantic segmentation in particular. Here, we incorporate metadata by employing a channel modulation mechanism in convolutional networks and study its effect on semantic segmentation tasks. We demonstrate that metadata as additional input to a convolutional network can improve segmentation results while being inexpensive in implementation as a nimble add-on to popular models. We hypothesize that this benefit of metadata can be attributed to facilitating multitask switching. This aspect of metadata-driven systems is explored and discussed in detail.

Keywords: semantic segmentation · metadata · multitasking

1 Introduction

Semantic segmentation is a technique broadly applied in biomedicine, separating critical regions according to their functional role in the source object, *e.g.*, detecting tumors from CT images [18] or differentiating cell nuclei from the background [2]. When performed by trained human experts, this task is time-consuming and costly [5]. The oldest automated techniques [4,19] in image processing were rule-based routines that separated regions on deterministic criteria. Since then, machine learning methods have taken over the field [8,15]. Most recently, deep learning models eliminated many manual steps in image processing and outperformed previous methods [10], but substantial improvements are still needed for robust systems.

Digital images often come with abundant metainformation reflecting image acquisition device settings, the methodology of sample preparation, or the

Supplementary Information The online version contains supplementary material available at https://doi.org/10.1007/978-3-031-45857-6_15.

provenance of the inspected objects. This information alone can say a lot about how images look and their quality, and potentially influence the segmentation results. For example, different cell lines have distinct appearances under the microscope [2], and knowing which types of cells to look for helps locate them.

Incorporating metadata into computer vision models is straightforward for the classification task. Kawahara *et al.* used a convolutional neural network (CNN) to extract features from skin images and concatenated them with a one-hot encoded metadata vector to analyze lesions [14]. Gessert *et al.* used an additional dense neural network to process age, anatomical site, and sex metadata and fused its features with the CNN [9]. However, mixing non-image data into the fully-convolutional neural network for semantic segmentation is less trivial. De Vries *et al.* introduced Conditional Batch Normalization to integrate additional information into feature maps [7]. This idea was developed into Feature-wise Linear Modulation (FiLM) [21] and Conditioned-U-Net [6]. Later, FiLM was adopted for the biomedical domain [17].

The idea of modulating CNN feature maps closely relates to channel attention, which uses learnable weights to emphasize relevant feature maps. Channel attention is actively employed to improve CNN's performance [26]. Hu *et al.* introduced a squeeze-and-excitation (SE) block to implement this idea by channel-wise feature recalibration [13]. Lightweight channel modulation blocks were beneficial for CNNs in various tasks [3,23].

This work explores the value of channel attention using SE block for metadata incorporation. We investigate the effects of categorical and continuous metadata on semantic segmentation and show that metadata improves the model's generalization and overall performance. We conduct experiments on biomedical datasets using metadata such as cell line labels and expected object size. Further, we explore the utility of the metadata to help the model effectively navigate multiple tasks on the same images. The results show a statistically significant advantage of metadata-driven models over metadata-free models. The difference becomes more pronounced when the model is trained on visually similar domains, and the system exhibits its multitasking properties. Overall, our contributions are as follows:

1. We present a novel, simple, lightweight, yet effective metadata-enhanced squeeze-and-excitation block.
2. We empirically show that using metadata improves the performance of the semantic segmentation models.
3. We find that metadata drastically increases the performance for the under-represented task in a multitask setting.

2   Methods

2.1   Metadata Incorporation with Channel Modulation

Squeeze-and-excitation (SE) blocks are a natural place for metadata fusion in a shallow bottleneck for channel recalibration. They consist of two linear layers, with ReLU activation for the first layer and sigmoid activation for the second.

Each block's input is squeezed from feature maps with pooling. The output is multiplied channel-wise with the same feature maps. We propose two ways of modifying an SE block. First, encoded metainformation can fully replace the squeezed input to the multi-layer perceptron. We call this block metadata-excitation (ME) because we abandon squeezing entirely. Second, we can concatenate metainformation to the squeezed vector allowing the network itself to decide on the importance of each input. We refer to this modification as squeeze-metadata-and-excitation (SME). Here, we test the performance of ME and SME blocks as components of U-Net network architecture against the baseline model with vanilla SE block.

We implement ME and SME blocks as a part of the modified U-Net [22] architecture (Fig. 1). The number of inputs depends on the configuration (number of feature maps, metadata elements, and selected type of block). Hidden dimensionality is four times lower than the number of feature maps.

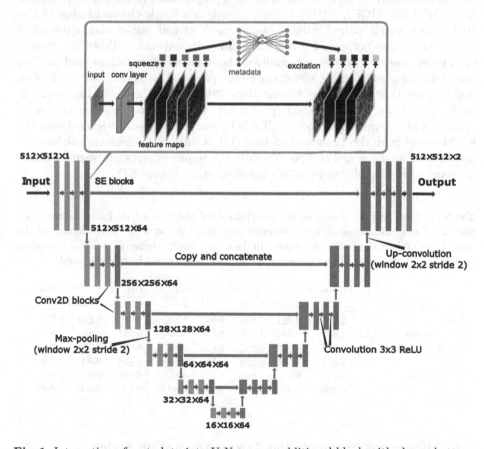

Fig. 1. Integration of metadata into U-Net as an additional block with channel attention functionality that can also receive encoded metadata affecting feature maps intensity. Here the schema for the SME model is shown. For the ME model squeeze part is absent, and the input to the linear layers consists of metadata only (red circles). (Color figure online)

2.2 Training

We used Adam optimizer [16] with CyclicLR [24] scheduler changing the learning rate linearly after each batch from 0.0002 to 0.0008 and back in 8 cycles for 100 epochs. We saved the best model by validation error for the downstream analysis. We implemented all the models in PyTorch 1.6 [20] and trained them on NVIDIA Tesla V100 16/32Gb GPUs with cuDNN 7.6.3.

3 Experiments

3.1 Microscopy Segmentation Across Cell Lines

To assess the impact of metadata on microscopy image segmentation, we trained our models on the Seven Cell Lines brightfield microscopy dataset [2]. It consists of 3024 samples from seven cell lines in equal proportions: HeLa, MDCK, HepG2, A549, HT1080, MCF7, NIH3T3. Each sample is a single-channel image of size 1080×1080 pixels paired with a binary mask of cell nuclei and a metadata vector with a one-hot encoded cell line label (Supplementary Table S1). Part of the models was trained conventionally without access to metadata, and another part had been assisted by metadata labels provided as additional input. Our best models that had access to metadata (SME) had shown an improvement in F_1 score from 0.861 to 0.863 ($p < 0.001$) compared to the baseline model trained without metainformation (Table 1). For comparison, we also used the FiLM model from the `ivadomed` toolbox [11]. It did not outperform the baseline model with an F1 score of 0.86. Visually the improvements are seen as better overlaps with ground truth masks (Supplementary Figure S1).

Table 1. Resulting F_1 scores of models trained on the Seven Cell Lines dataset. See Sect. 2.1 for method description. "baseline stratified" denotes the performance of the models trained on the particular subset of data (cell line), "dummy" denotes metadata vector filled with zeros. The best score is in **bold**, the second best underlined.

	Average	HeLa	MDCK	A549	HT1080	HepG2	MCF7	NIH3T3
baseline stratified	-	0.901	0.830	0.841	0.857	0.806	0.825	0.891
baseline	0.861	0.901	0.865	0.852	0.870	0.806	0.840	0.891
FiLM-dummy [6, 17, 21]	0.855	0.895	0.858	0.846	0.864	0.802	0.835	0.885
FiLM [6, 17, 21]	0.860	0.900	0.865	0.853	0.867	0.806	0.839	0.890
ME	0.861	0.900	0.865	0.852	0.869	0.808	**0.842**	0.891
SME-dummy	0.859	0.899	0.864	0.852	0.867	0.805	0.838	0.889
SME	**0.863**	**0.902**	**0.869**	**0.854**	**0.872**	**0.809**	**0.842**	**0.893**

3.2 Microscopy Segmentation Across Annotation Styles

Microscopists can exhibit different annotation styles when drawing an object boundary as polygon vertices. More points provide more accurate segmentation, while fewer points save labeling time. From the Seven Cell Lines training dataset we derived a dataset with three nuclei stencil styles using various polygonization degrees of `skimage.measure.approximate_polygon()` [25]. One-third of masks were left unchanged. The second third had a fine approximation to polygons. The remaining part had coarser annotations (Supplementary Figure S2). The test and validation parts had accurate masks and remained consistent with the Seven Cell Lines dataset. Thus, we aimed to obtain fine-grained segmentation results with coarsely annotated parts of the training set.

Here, a metadata label serves as a task identifier, making the model predict masks with shape and confidence specific to the annotation style. The coarse style lowers the prediction confidence, affecting the area of the binarized mask. Switching the model to the mode with accurate masks during the evaluation boosts confidence and helps increase accuracy on the test set (Supplementary Figure S3). All models with metadata had higher F_1 scores, with improvements from 0.846 to 0.854 with our best SME model compared to the baseline model trained on all three parts of the dataset without metadata ($p < 0.001$, Table 2).

Table 2. Results of the experiment with the dataset with different annotation styles. Row "baseline 1/3" denotes results from the SE model trained on one-third of data with accurate masks. The main baseline model was trained on the full dataset without metadata. ME, SME, and FiLM models use metadata with different implementations. SME dummy - SME model with meaningless metadata (zeros in the input vector). The best score is in **bold**, the second best underlined.

	F_1 score
baseline 1/3 data	0.837
baseline	0.846
ME	<u>0.850</u>
SME	**0.854**
FiLM	0.849
SME dummy	0.848

3.3   Microscopy Multilabel Segmentation: Cells and Anomalies

Researchers prefer a clean dataset, as visual artifacts hinder downstream analysis [1]. However, anomaly segmentation for microscopy remains a challenging task because of the lack of annotations and their diverse shapes and sizes. We propose to take advantage of nuclei mask abundance and use metadata as a switch for multitask learning. Thus, we train the network to produce nuclei and anomaly segmentation from a single head, depending on the metadata input. To assess this approach and compare it with a conventional parallel multi-headed prediction, we expand the Seven Cell Lines dataset with multilabel targets. For 365 images out of 3024, we additionally present anomaly segmentation masks, highlighting debris and optical defects. For these images, metadata encodes the mask type: nuclei or anomaly. Since anomaly annotations are absent for the rest of the dataset, we are able to train parallel multi-headed networks only on this subset with two segmentation masks. On the contrary, we can train on the whole dataset when using metadata as a task switch.

A metadata-driven model achieved an F_1 score of 0.836 and 0.85 for nuclei and anomalies, respectively (Table 3). It notably outperformed the baseline model trained to segment only anomalies that had F_1 score of 0.736 on the anomaly subset. At the same time, the performance for nuclei segmentation slightly dropped from the baseline F_1 score of 0.854 on the nuclei subset. Expectedly, swapping metadata labels resulted in segmentation failure, approaching zero F1 score. The model with dummy metadata input also struggled with anomaly segmentation due to the dominance of nuclei masks, predicting only nuclei.

Additionally, we utilized a limited version of the large dataset containing nuclei and anomalies on the same images. Training a multi-headed model parallelly on this reduced dataset, with each head's output corresponding to a different mask, yielded a higher F_1 score of 0.835 on the anomaly subset than the baseline model with F_1 score of 0.736 training exclusively on that subset. The performance on the nuclei mask subset was lower due to the limited training size (the stratified baseline model trained on a limited subset of nuclei was not introduced in this experiment due to the focus on anomaly segmentation).

Surprisingly, when our initial metadata-driven model with a single output was retrained on the reduced dataset in the usual sequential mode using metadata to switch segmentation task, the anomaly subset showed the highest boost in F_1 score among all experiments: 0.854 compared to the individual model with F_1 score of 0.736. Detailed results are summarized in Table 3.

Table 3. F_1 scores for multitask models from experiments on dataset with anomalies (full and multilabel subset). See Sect. 2.1 for method description. "baseline stratified" denotes the performance of the SE models trained on the particular subset of data (anomaly/nuclei masks), "dummy" denotes the model with metadata vector filled with zeros. The best score is in **bold**, the second best underlined.

		Average	Anomalies	Seven Cell Lines
Full dataset	baseline stratified	-	0.736	0.854
	ME	0.835	0.824	0.845
	SME-dummy	0.470	0.104	0.835
	SME	0.843	<u>0.836</u>	0.850
Multilabel subset	Two heads	0.812	0.835	0.789
	SME	0.823	**0.854**	0.792

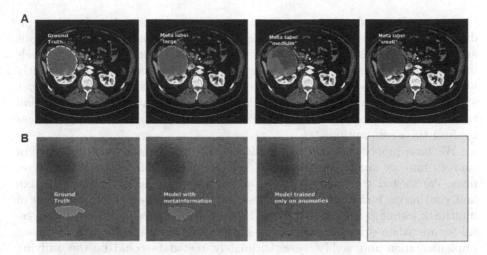

Fig. 2. (A) Manifestation of multitask properties in metadata-driven models with continuous metadata correlated with the tumor size on KiTS21 dataset. In this experiment, the baseline model had on the test F_1 score of 0.734, and metadata-driven models (SME and ME) had an average F_1 score of 0.804 for two models. (B) Example of positive transfer for the anomaly segmentation task with the metadata-driven model — the baseline model (trained only on the anomaly subset) had F_1 score of 0.736 on this subset, metadata-driven models (SME) had F_1 score of 0.854 on anomaly subset.

3.4   Kidney Tumor Segmentation with Continuous Metadata

To test our method's applicability for continuous metadata, we used KiTS21 [12] 3D dataset of 200 images accompanied with the tumor size information. We derived a 2D dataset from it with 10346 images selecting axial slices containing kidneys. Segmentation targets contain three classes: kidney tissue, tumor, and cyst (cyst and tumors were exclusive classes — cases with tumors did not have cysts and vice versa). In this experiment, the baseline model had a F_1 score of 0.734 on the test, metadata-driven model ME achieved a F_1 score of 0.807, and another metadata-driven SME model achieved a F_1 score of 0.801. In Fig. 2A, we illustrate how the target mask changes in response to different size-correlated metadata applied to the same image.

4   Discussion

Metadata is a widely available and potentially valuable source of additional information for deep learning systems. We used a channel modulation mechanism to inject it into the image-processing workflow, a lightweight and low-intrusive approach for adapting existing CNNs. Despite small dimensionality compared to the main input, metadata can drastically change the course of learning and inference by indicating the presence of a specific task. Our experiments show how this change is especially obvious when no other visual clues are present in the image. The results demonstrate that the CNN models endowed with meta-data awareness respond strongly to the metadata input, adjusting their output according to the respective task indicator. While the correct metadata labels improve the segmentation performance, providing incorrect labels significantly hinders the model's functionality. The metadata-driven system inherently has a multitasking nature that allows exploiting the benefits of combining several tasks. Our results for segmenting anomalies and nuclei from microscopy images confirm these assumptions.

We have proposed a method of incorporating metadata from a variety of sources into an image segmentation pipeline and demonstrated its effective-ness. We showed that metadata can act as a guiding label (in both discrete and continuous form) or even as a task switch. We believe that principles of multitask learning, some of which we have uncovered here, will remain univer-sal for metadata-driven segmentation systems and persist regardless of chosen implementation and architecture. Ultimately, metadata could be the unifying modality that allows training truly cross-domain models.

Acknowledgements. This work was funded by Revvity, Inc. (previously known as PerkinElmer Inc., VLTAT19682), and Wellcome Trust (206194). We thank High Per-formance Computing Center of the Institute of Computer Science at the University of Tartu for the provided computing power.

References

1. Ali, M.A.S., et al.: ArtSeg: rapid artifact segmentation and removal in brightfield cell microscopy images, January 2022
2. Ali, M.A.S., Misko, O., Salumaa, S.O., Papkov, M., Palo, K., Fishman, D., Parts, L.: Evaluating very deep convolutional neural networks for nucleus segmentation from brightfield cell microscopy images. SLAS Discov. **26**(9), 1125–1137 (2021)
3. Amer, A., Ye, X., Zolgharni, M., Janan, F.: ResDUnet: residual dilated UNet for left ventricle segmentation from echocardiographic images. Conf. Proc. IEEE Eng. Med. Biol. Soc. **2020**, 2019–2022 (2020)
4. Beucher, S.: Use of watersheds in contour detection. In: Proceedings of Interna-tional Workshop on Image Processing, Sept. 1979, pp. 17–21 (1979)
5. Bhalgat, Y., Shah, M., Awate, S.: Annotation-cost minimization for medical image segmentation using suggestive mixed supervision fully convolutional networks, December 2018
6. Brocal, G.M., Peeters, G.: Conditioned-U-Net: introducing a control mechanism in the U-Net for multiple source separations. In: Proceedings of the 20th International Society for Music Information Retrieval Conference. Zenodo (2019)
7. De Vries, H., Strub, F., Mary, J., Larochelle, H., Pietquin, O., Courville, A.C.: Modulating early visual processing by language. Adv. Neural Inf. Process. Syst. **30** (2017)
8. Georgiou, T., Liu, Y., Chen, W., Lew, M.: A survey of traditional and deep learning-based feature descriptors for high dimensional data in computer vision. Int. J. Multimed. Inf. Retrieval **9**(3), 135–170 (2020)
9. Gessert, N., Nielsen, M., Shaikh, M., Werner, R., Schlaefer, A.: Skin lesion classifi-cation using ensembles of multi-resolution EfficientNets with meta data. MethodsX **7**, 100864 (2020)

10. Greenspan, H., van Ginneken, B., Summers, R.M.: Guest editorial deep learning in medical imaging: Overview and future promise of an exciting new technique. IEEE Trans. Med. Imaging **35**(5), 1153–1159 (2016)
11. Gros, C., Lemay, A., Vincent, O., Rouhier, L., Bourget, M.H., Bucquet, A., Cohen, J., Cohen-Adad, J.: Ivadomed: a medical imaging deep learning toolbox. J. Open Source Softw. **6**(58), 2868 (2021)
12. Heller, N., et al.: The state of the art in kidney and kidney tumor segmentation in contrast-enhanced CT imaging: results of the KiTS19 challenge. Med. Image Anal. **67**, 101821 (2021)
13. Hu, J., Shen, L., Albanie, S., Sun, G., Wu, E.: Squeeze-and-excitation networks. IEEE Trans. Pattern Anal. Mach. Intell. **42**(8), 2011–2023 (2020)
14. Kawahara, J., Daneshvar, S., Argenziano, G., Hamarneh, G.: 7-point checklist and skin lesion classification using Multi-Task Multi-Modal neural nets. IEEE J. Biomed. Health Inform., April 2018
15. Khan, S., Sajjad, M., Hussain, T., Ullah, A., Imran, A.S.: A review on traditional machine learning and deep learning models for WBCs classification in blood smear images. IEEE Access **9**, 10657–10673 (2021)
16. Kingma, D.P., Ba, J.: Adam: a method for stochastic optimization, December 2014
17. Lemay, A., Gros, C., Vincent, O., Liu, Y., Cohen, J.P., Cohen-Adad, J.: Benefits of linear conditioning for segmentation using metadata. In: Heinrich, M., Dou, Q., de Bruijne, M., Lellmann, J., Schläfer, A., Ernst, F. (eds.) Proceedings of the Fourth Conference on Medical Imaging with Deep Learning. Proceedings of Machine Learning Research, vol. 143, pp. 416–430. PMLR (2021)
18. Litjens, G., et al.: A survey on deep learning in medical image analysis. Med. Image Anal. **42**, 60–88 (2017)
19. Otsu, N.: A threshold selection method from gray-level histograms. IEEE Trans. Syst. Man Cybern. **9**(1), 62–66 (1979)
20. Paszke, A., et al.: Pytorch: an imperative style, high-performance deep learning library. In: Advances in Neural Information Processing Systems, pp. 8026–8037 (2019)
21. Perez, E., Strub, F., de Vries, H., Dumoulin, V., Courville, A.: FiLM: Visual reasoning with a general conditioning layer. AAAI **32**(1), April 2018
22. Ronneberger, O., Fischer, P., Brox, T.: U-Net: convolutional networks for biomedical image segmentation. In: Navab, N., Hornegger, J., Wells, W.M., Frangi, A.F. (eds.) MICCAI 2015. LNCS, vol. 9351, pp. 234–241. Springer, Cham (2015). https://doi.org/10.1007/978-3-319-24574-4_28
23. Roy, S.K., Dubey, S.R., Chatterjee, S., Chaudhuri, B.B.: FuSENet: fused squeeze-and-excitation network for spectral-spatial hyperspectral image classification (2020)
24. Smith, L.N.: Cyclical learning rates for training neural networks. In: 2017 IEEE Winter Conference on Applications of Computer Vision (WACV), pp. 464–472, March 2017
25. Van der Walt, S., Schönberger, J.L., Nunez-Iglesias, J., Boulogne, F., Warner, J.D., Yager, N., Gouillart, E., Yu, T.: scikit-image: image processing in python. PeerJ **2**, e453 (2014)
26. Woo, S., Park, J., Lee, J.-Y., Kweon, I.S.: CBAM: convolutional block attention module. In: Ferrari, V., Hebert, M., Sminchisescu, C., Weiss, Y. (eds.) ECCV 2018. LNCS, vol. 11211, pp. 3–19. Springer, Cham (2018). https://doi.org/10.1007/978-3-030-01234-2_1

Self-prompting Large Vision Models for Few-Shot Medical Image Segmentation

Qi Wu[1(✉)], Yuyao Zhang[1], and Marawan Elbatel[1,2]

[1] The Hong Kong University of Science and Technology, Hong Kong, China
{qwuaz,yzhangkp}@connect.ust.hk
[2] Computer Vision and Robotics Institute, University of Girona, Girona, Spain

Abstract. Recent advancements in large foundation models have shown promising potential in the medical industry due to their flexible prompting capability. One such model, the Segment Anything Model (SAM), a prompt-driven segmentation model, has shown remarkable performance improvements, surpassing state-of-the-art approaches in medical image segmentation. However, existing methods primarily rely on tuning strategies that require extensive data or prior prompts tailored to the specific task, making it particularly challenging when only a limited number of data samples are available. In this paper, we propose a novel perspective on self-prompting in medical vision applications. Specifically, we harness the embedding space of SAM to prompt itself through a simple yet effective linear pixel-wise classifier. By preserving the encoding capabilities of the large model, the contextual information from its decoder, and leveraging its interactive promptability, we achieve competitive results on multiple datasets (i.e. improvement of more than 15% compared to fine-tuning the mask decoder using a few images). Our code is available at https://github.com/PeterYYZhang/few-shot-self-prompt-SAM

Keywords: Image Segmentation · Few-shot Learning · SAM

1 Introduction

Supervised methods in medical image analysis require significant amounts of labeled data for training, which can be costly and impractical due to the scarcity of high-quality labeled medical data. To solve this problem, many works like [1,24], adopted few-shot learning, which aims at generalizing model to a new class via learning from a small number of samples. However, such methods tried to learn contextual information of the new class, which is hard since the contextual information can be complex and multi-faceted, and can be easily influenced by noise. Therefore, we seek for methods that require less information, such as the size and location of the segmentation target.

Recent advancements in large-scale models, such as GPT-4 [20], DALL-E [22], and SAM [14], have shed light on few-shot and even zero-shot learning. Due to

---

Q. Wu and Y. Zhang—Co-first authors.

L. Koch et al. (Eds.): DART 2023, LNCS 14293, pp. 156–167, 2024.
https://doi.org/10.1007/978-3-031-45857-6_16

their remarkable capabilities in transferring to multiple downstream tasks with limited training data, these models can act as foundation models with exceptional generalization abilities, and prompts play a crucial role in determining their overall performances. Many downstream tasks benefit from these models by leveraging prompt engineering [6,30] and fine-tuning techniques [16,28,29]

One prominent large foundation model in computer vision, the Segment Anything Model (SAM) [14], is a powerful tool for various segmentation tasks, trained on natural images. The model can generate different masks based on different user input prompts. Due to SAM's promptable nature, it can potentially assist medical professionals in interactive segmentation tasks.

When solving practical tasks in clinics and hospitals, several challenges need to be addressed: 1) How to tackle the scarcity of medical data, and 2) how to be user-friendly and assist medical professionals in more flexible way.

However, though typical few-shot learning models can reduce the data required, they do not have the promptable feature. Also, while some other SAM fine-tuning method [16,29] can achieve promptability, more labelled data are required during training.

Recently, self-prompting arises in tuning large language models (LLMs) [15], where the model prompts itself to improve the performance. To overcome the two aforementioned challenges simultaneously, we draw inspiration from the success of self-prompting LLMs and propose a novel method that utilizes a simple linear pixel-wise classifier to self-prompt the SAM model. Our method leverage the promptable feature of large vision foundation model, having a simple architecture by inserting a small plug-and-play unit in SAM. At the same time, all the training can be done with limited labelled data and time. Remarkably, our method can already achieve good results using only a few images training set, outperforming some other fine-tuning methods [16,29] that use the same amount of data. Furthermore, our method is almost training-free, the training can be done within 30 s while other fine-tuning methods require more than 30 min (in the few-shot setting). This allows generation of output masks with few computational resources and time, which can assist medical professionals in generating more precise prompts or labeling data.

To summarize, our major contributions are:

- We propose a novel computational efficient method that leverages the large-scaled pre-trained model SAM for few-shot medical image segmentation
- We develop a method to self-prompt the foundation model SAM in the few-shot setting and demonstrated the potential and feasibility of such self-prompting method for medical image segmentation
- We experiment and show that our method outperforms other SAM fine-tuning methods in a few-shot setting and is more practical in clinical use

2   Related Works

2.1   Few-Shot Medical Image Segmentation

Few-shot learning has been popular in medical image segmentation, as it requires significantly less data while still reaching satisfactory results. Previous works [1,7,17,24,25,27] have shown great capacity for few-shot learning in some different medical segmentation tasks. But this methods aimed to learn the prototype knowledge or contextual information of the target domain, which is easily influenced by noise and other factors since those information are complex and multi-faceted. Unlike these methods, we propose a novel technique that utilizing a large vision foundation model to achieve few-shot learning.

2.2   SAM

Inspired by the "prompting" techniques of NLP's foundation models [4], the project team defined the segment anything task as returning valid segmentation masks given any segmentation prompts. To introduce zero-shot generalization in segmentation, the team proposed the Segment Anything Model (SAM) [14], which is a large-scaled vision foundation model. Several recent studies, including [3,8,9,12,18,19,30], have evaluated SAM's capability on different medical image segmentation tasks in the context of zero-shot transfer. The results show that SAM can generate satisfactory masks with sufficient high-quality prompts for certain datasets. However, manually and accurately prompting SAM from scratch will be time-consuming and inefficient for medical professionals. Our method can self-generate prompts, hence being more user-friendly and can assist professionals during inference.

2.3   Tuning the Segment Anything Model

Large Foundation models in vision including SAM are difficult to be trained or tuned from scratch due to the limitation of computing resources and data. Multiple previous works [16,28,29] have fine-tuned SAM on medical datasets. Specifically, MedSAM [16] fine-tune the SAM mask decoder on a large-scaled datasets, SAMed [29] adopt low-rank-based fine-tuning strategy (LoRA) [10] and train a default prompt for all image in the dataset, Medical SAM Adapter (MSA) [28] use adapter modules for fine-tuning. These methods yield satisfactory results, getting close to or even outperforming SOTA fully-supervised models. However, these SAM-based works still needs large amounts of data to fine-tune the model in a supervised way, yet have not fully leverage the prompting ability.

3   Methodology

We denote the training dataset as $D = \{I, T\}$, where $I = \{i_1, \ldots, i_n\}$, and $T = \{t_1, \ldots, t_n\}$, $n \in N$ corresponding to the images and segmentation ground truth.

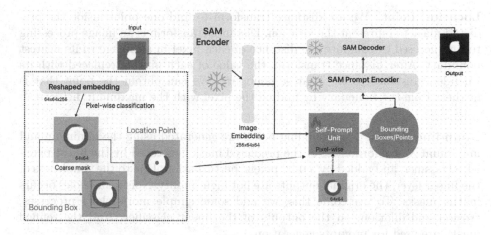

Fig. 1. The overall design of our framework. The pink modules are exactly the same as the original SAM architecture, and the "snowflake" sign represents that we freeze the module during training. The Self-Prompt Unit ("fire" sign) is trained using the image embeddings from the SAM encoder and the resized ground truth label. The unit predicts a coarse mask, which is used to obtain the bounding box and location point that prompt SAM. (Color figure online)

Our goal is to design an plug-and-play self-prompting unit that can provide SAM with the **location** and **size** information of the segmentation target with only a **few labeled data**, k images for example, denoted as $D_k = \{I_k, T_k\}$. As shown in Fig. 1, our model build upon the original SAM model(the pink blocks) which is kept frozen all the time during training and inference. For each image i_n, the image encoder maps it to the embedding spaces $z_n = E_{SAM}(i_n)$, then our self-prompt unit take the image embedding to provide the bounding box and point as prompt p_n. Finally, the decoder combined the encoded prompt $E_{prompt}(p_n)$ and the image embedding z_n to get the final segmentation output.

3.1 Self-prompt Unit

To learn the location and size information of the target, an intuitive way is to get a coarse mask as a reference. After passing through the powerful encoder of SAM, which is a Vision Transformer (ViT) [5], the input image is encoded as a vector $z_n \in R^{256 \times 64 \times 64}$. To align the mask with the encoded image embedding, we down-sample it to 64×64. Here the mask is treated as binary, then we conduct a logistic regression to classify each pixel as background or mask to get the coarse mask. Also, using a logistic regression instead of neural networks will minimize the influence of inference speed. Finally, from the predicted low-resolution mask, the location point and the bounding box can be obtained using morphology and image processing techniques.

Location Point. We use distance transform to find one point inside the predicted mask to represent the location. Distance transform is an image processing technique used to compute the distance of each pixel in an image to its nearest boundary. After distance transform, the value of each pixel is replaced with its Euclidean distance to the nearest boundary. We can obtain the point that is farthest from the boundary by finding the pixel with the maximum distance.

Bounding Boxes. The bounding boxes are generated using the minimum and maximum X, Y coordinates of the predicted mask generated by the linear pixel-wise classifier, and added by a 0-20 pixels' perturbation. Due to the simpleness of the linear layer, the original outputs are not high-quality. Noises and holes occurs in the masks. To overcome this, we add some simple morphology processes, erosion and dilation, on the outputs of the linear classifier. And the refined masks are used for prompts generation.

3.2   Training Objectives

Each image-mask pair in the training dataset $D_k = \{I_k, T_k\}$ is denoted as $\{i^q, t^q\}$, for simplicity we denote the down-sampled mask the same as the original one. The image i^q is first fed to the image encoder to get the image embedding $z^q \in R^{256 \times 64 \times 64}$ then is reshaped to $z^q \in R^{64 \times 64 \times 256}$ in corresponding to the mask of shape $t^q \in R^{64 \times 64}$. Since we perform the Logistic Regression pixel-wisely, the loss function becomes,

$$L = \frac{1}{k} \sum_q^k \sum_{1 \le m,n \le 64} -(t^q_{m,n} \log \hat{t}^q_{m,n} + (1 - t^q_{m,n} \log(1 - \hat{t}^q_{m,n}))) \quad (1)$$

where, $t^q_{m,n}$ is the value of pixel (m, n) of the q-th mask in the subset D_k, and $\hat{t}^q_{m,n} = \mathbf{1}_{\sigma(w^T z^q_{m,n} + b) > 0.5}$, where $z^q_{m,n}, w, b \in R^{256}$, and σ is the Sigmoid function.

4   Experiments

4.1   Datasets

We split the datasets into 5 parts with equal size and apply k-fold (k=5) cross-validation to evaluate our method.

Kvasir-SEG. The Kvasir-SEG dataset [11] contains 1000 polyp images and their corresponding ground truth. It has JPEG encoding and includes bounding box coordinates in a JSON file. The task on this dataset is to segment polyps.

ISIC-2018. The ISIC-2018 dataset [2,26] was published by the International Skin Imaging Collaboration (ISIC) as a comprehensive collection of dermoscopy images. It is an aggregate dataset that comprises a total of 2594 images, providing a diverse range of examples for analysis. This dataset is for skin lesion segmentation.

4.2 Implementation Details

We use the ViT-B SAM model, the smallest version of SAM, as our backbone.

First, we use the SAM image encoder (ViT-B) to obtain the image embeddings. In the embedding space, the spatial resolution is 64×64 pixels and each pixel corresponds to a 256-dimensional vector that represents the features extracted from the original image. To train a pixel-wise classifier, we utilize these pixel-wise embeddings alongside the ground truth labels. The ground truth labels are resized to 64×64 to match the resolution of the embeddings. Each pixel in this resized label image corresponds to a specific class.

For simplicity, we utilize the logistic regression module directly from the scikit-learn [21] library. We set the maximum iteration to 1000 and use the default values for all other hyper-parameters. We make this decision because we are only interested in generating simple prompts and believe that intricate hyper-parameter tuning would be unnecessary and will not significantly improve our results. By doing this, we can focus on creating and testing our prompts without tuning the hyper-parameters for different datasets.

During the inference phase, the linear pixel-wise classifier takes an image embedding as input and produces a low-resolution mask of size 64×64. This mask is then resized to 256×256 and subjected to a 3-iteration erosion followed by a 5-iteration dilation using a 5×5 kernel. The resulting mask is used to generate prompts, including a point and a bounding box, for further processing. The prompts, together with the image embeddings, are then passed to the SAM mask decoder to generate the final mask.

4.3 Results

We use Dice and IoU as our evaluation metrics. We compare our method with another two fine-tuning models, MedSAM [16] and SAMed [29]. We also include the result of the original unprompted SAM-B by extracting the masks that best overlap the evaluated ground truth (same as the setting in [30]). Notice that we did not utilize the ground truth in any mask extraction for evaluating our method. For MedSAM, we set the bounding box (xmin, ymin, xmax, ymax) as the image size (0, 0, H, W) for all images, and trained the decoder. For SAMed, we only fine-tuned the decoder for fairness. We set the size of the training set to be 20. The reason for 20 shots instead of fewer shots is that, it is hard for the other two fine-tuning methods to generate valid masks with such limited training data. We want to obtain a more valuable comparison. The performance of our approach using less number of shots remains competitive, which we demonstrate in the ablation study.

The results are shown in Table 1 and some examples are in Fig. 2. Our method using both the point and bounding box reaches 62.78% and 53.36% in Dice and IoU score on Kvasir-SEG and 66.78% and 55.32% on ISIC2018 correspondingly, which surpasses MedSAM [16], SAMed [29] and the original unprompted SAM. Notice that training full data on ISIC2018, the metrics are a lot higher than our methods trained on 20 images, this may because the dataset is quite large, more than 2000 images. More discussion of our results will be presented in ablation study.

Remarkably, the training of our method can be done in 30 s using an NVIDIA RTX-3090 GPU. In fact, most of the time is used for computing image embeddings. If the embeddings are pre-computed, the training can be done within a few seconds using a CPU. In comparison, for other fine-tuning methods, at least 30 min are required in the few-shot setting. Some limitations of our work are discussed in the supplementary.

Table 1. Comparison of our method (using both of the bounding box and the point) to other fine-tuning methods in a few-shot setting. Number of shots is set to be 20 here. We use the ground truth to extract the best masks for unprompted SAM, the prompted one is an upper bound in which prompts are generated from the ground truth. The results of other methods fine-tuned on the whole dataset are also listed for reference. Notice that, for SAMed, we only trained the decoder. The U-net here is pre-trained on ImageNet.

Models	Kvasir-SEG		ISIC2018	
	Dice%↑	IoU%↑	Dice%↑	IoU%↑
Ours	**62.78**	**53.36**	**66.78**	**55.32**
MedSAM [16]	55.01	43.21	64.94	54.65
SAMed [29]	61.48	51.75	63.27	54.23
Unprompted SAM-B [30]	52.66	44.27	45.25	36.43
MedSAM [16] (full-data)	66.14	55.77	84.22	75.49
SAMed [29] (full-data)	79.99	71.07	85.49	77.40
Prompted SAM-B	86.86	79.49	84.28	73.93
U-net [23] (full-data)	88.10	81.43	88.36	81.14

5   Ablation Study

5.1   Number of Shots

We perform ablation study on the number of shots (k) of our model. We choose $k = 10, 20, 40$ and full dataset. The outcomes obtained on ISIC-2018 are presented in Table 2, whereas the results of Kvasir-SEG are provided in the supplementary materials. The overall performance is improved with more shots. However, the improvement is limited. This is understandable since we only train a simple logistic regression. Due to the limited performance of the classifier, generating more valid prompts is hard. But if we train a more sophisticated classifier, it will like an decoder and deviate from our main goal of reducing data and computation requirement. Iterative use of masks and more advanced prompt generating mechanism can be explored in the future.

5.2 Use of Methods

We also conduct the ablation study of the methods: 1) using only the point; 2) using only the bounding box; 3) using both the point and the bounding box. Results are shown in Table 2. The bounding box provides the size information and a rough position of the instance, while the point gives the accurate position of the instance. Figure 2 demonstrates some examples of utilizing different prompt generation approaches. Since we get the coarse mask from the linear pixel-wise classifier, the size information of the instance may be highly influenced by the mistakenly classified pixels (i.e. the left example), in this case, points will help to locate the object. But the point itself cannot provide size information, which may cause the model to generate masks with wrong size (i.e. the right example), in this case boxes will help to restrict the size. Taking the advantages of both the point and the bounding box, we can obtain better results. More results, including failure cases, are included in our GitHub repository.

Table 2. Results of the ablation study showing the impact of number of shots and different prompting methods. "Linear" means the coarse mask from the linear pixel-wise classifier; "point" means only using the point as prompt; "box" represents only using the bounding box as prompt; "point+box" means using both of bounding box and point as prompt. The scores here are dice scores.

ISIC-2018				
	10 shots	20 shots	40 shots	full-data
Ours(linear)	58.81	62.69	63.81	65.62
Ours(point)	59.49	61.23	61.52	61.76
Ours(box)	47.13	47.99	49.81	55.03
Ours(point+box)	64.22	66.78	67.88	69.51

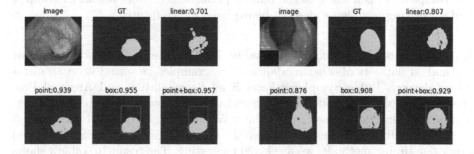

Fig. 2. Some examples of using different methods on Kvasir-SEG dataset. The yellow objects in the figures denote the segmented polyps. "Linear" here is the coarse mask from the linear pixel-wise classifier. The scores here are Dice scores. (Color figure online)

6   Conclusion

Our study has demonstrated the potential and feasibility of utilizing self-prompting with large-scale vision foundation models for medical image segmentation. We present a simple yet effective idea, using a few images to train a linear pixel-wise classifier on the image embedding space to generate a prompt for SAM. Our method can be more user-friendly than traditional few-shot learning models due to SAM's promptable feature and requires significantly less data than those SAM fine-tuning models. We evaluate our method on two datasets. The results show that our method outperforms another two fine-tuning methods using same amount of data. Since the whole process requires few computational resources and time, the resulting outputs can also be regarded as raw masks that can assist medical professionals in more precise prompt generation and data labeling. Future studies can lay more emphasis on getting more accurate prompts from the output of SAM. Combining self-prompting with other fine-tuning methods can also be explored.

Acknowledgement. M.E is partially funded by the EACEA Erasmus Mundus grant. We would like to acknowledge Prof. Xiaomeng Li for revising our manuscript. We would also like to thank Mr. Haonan Wang for providing valuable suggestions to our work.

A   Supplementary Materials

A.1   Limitation

Multi-instance Segmentation. The first limitation of our method lies in the segmentation task that has multiple instances. One can blame the problem on the plugged-in linear classifier, the simple classifier cannot know the number of instances accurately, so the spatial information passed to the prompt encoder is not complete, thus leading to the limited performance. More advanced training and prompting techniques need to be explored in the future.

Limitation of Modality Knowledge in Decoder. We also tested our method on datasets of other modalities. For example, we tested it on an ultrasound dataset, Pubic Symphysis-Fetal Head Segmentation and Angle of Progression [13], which includes lots of high-frequency features. We test head segmentation using this dataset, see Table 3 We found that the SAM decoder will lead to degrading performance. To keep the fairness for both of our methods and other fine-tuning methods, we use k=20 for testing. The result in Table 3 shows that our method has a lower performance compared to MedSAM and SAMed. Surprisingly, in the examples in Fig. 3, we found that the performance is better when just using the linear classifier and then upscaling. The reason is that the decoder of SAM does not have the capability to predict the accurate mask from the interference of high-frequency features. This reflects that although the size and position of the instance are important, the classifier needs to know the basic knowledge of the modality. To solve it, one may combine our methods together with other fine-tuned decoder.

Table 3. The results of our method on the ultrasound dataset: Symphysis-Fetel, in a 20-shot setting. "Linear" means the coarse mask generated by the linear pixel-wise classifier, "point+box" means both of the self-generated point and box are used for the final output.

Models	Symphysis-Fetal	
	Dice	IoU
Ours(linear)	66.47	53.32
Ours(point+box)	69.67	55.94
MedSAM [16]	73.78	61.22

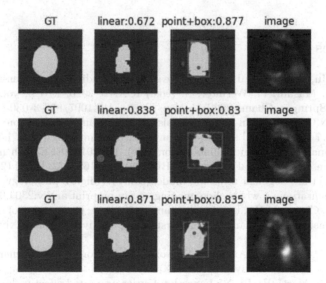

Fig. 3. Some examples of the result of our method on the Pubic Symphysis-Fetel dataset. The segmentation result is not satisfactory in ultrasound images, although the score is high. Also, the linear classifier even outperforms our method in some case. The result show that original SAM is sensitive to high-frequency perturbations (i.e. edges or noise in ultrasound).

A.2   Ablation Study Table

Table 4. Results of the ablation study on Kvasir-SEG. "Linear" means the coarse mask from the linear pixel-wise classifier; "point" means only using the point as prompt; "box" represents only using the bounding box as prompt; "point+box" means using both of bounding box and point as prompt. The scores here are dice scores.

Kvasir-SEG				
	10 shots	20 shots	40 shots	full-data
Ours(linear)	50.42	49.46	52.91	54.32
Ours(point)	58.50	62.51	63.24	63.51
Ours(box)	46.55	51.95	54.91	58.12
Ours(point+box)	60.03	62.78	65.34	67.08

References

1. Cai, A., Hu, W., Zheng, J.: Few-shot learning for medical image classification. In: Farkaš, I., Masulli, P., Wermter, S. (eds.) ICANN 2020. LNCS, vol. 12396, pp. 441–452. Springer, Cham (2020). https://doi.org/10.1007/978-3-030-61609-0_35

2. Codella, N.C., et al.: Skin lesion analysis toward melanoma detection: A challenge at the 2017 international symposium on biomedical imaging (isbi), hosted by the international skin imaging collaboration (isic). In: 2018 IEEE 15th international symposium on biomedical imaging (ISBI 2018), pp. 168–172. IEEE (2018)

3. Deng, R., et al.: Segment anything model (sam) for digital pathology: assess zero-shot segmentation on whole slide imaging. arXiv preprint arXiv:2304.04155 (2023)

4. Devlin, J., Chang, M.W., Lee, K., Toutanova, K.: Bert: pre-training of deep bidirectional transformers for language understanding. arXiv preprint arXiv:1810.04805 (2018)

5. Dosovitskiy, A., et al.: An image is worth 16x16 words: transformers for image recognition at scale. arXiv preprint arXiv:2010.11929 (2020)

6. Elbatel, M., Martí, R., Li, X.: Fopro-kd: fourier prompted effective knowledge distillation for long-tailed medical image recognition. ArXiv abs/ arXiv: 2305.17421 (2023)

7. Feyjie, A.R., Azad, R., Pedersoli, M., Kauffman, C., Ayed, I.B., Dolz, J.: Semi-supervised few-shot learning for medical image segmentation. arXiv preprint arXiv:2003.08462 (2020)

8. He, S., Bao, R., Li, J., Grant, P.E., Ou, Y.: Accuracy of segment-anything model (sam) in medical image segmentation tasks. arXiv preprint arXiv:2304.09324 (2023)

9. Hu, C., Li, X.: When sam meets medical images: An investigation of segment anything model (sam) on multi-phase liver tumor segmentation. arXiv preprint arXiv:2304.08506 (2023)

10. Hu, E.J., et al.: Lora: low-rank adaptation of large language models. arXiv preprint arXiv:2106.09685 (2021)

11. Jha, D., et al.: Kvasir-SEG: a segmented polyp dataset. In: Ro, Y.M., et al. (eds.) MMM 2020. LNCS, vol. 11962, pp. 451–462. Springer, Cham (2020). https://doi.org/10.1007/978-3-030-37734-2_37

12. Ji, W., Li, J., Bi, Q., Li, W., Cheng, L.: Segment anything is not always perfect: an investigation of sam on different real-world applications. arXiv preprint arXiv:2304.05750 (2023)

13. Jieyun, B.: Pubic Symphysis-Fetal Head Segmentation and Angle of Progression (Apr 2023). https://doi.org/10.5281/zenodo.7851339, https://doi.org/10.5281/zenodo.7851339

14. Kirillov, A., et al.: Segment anything. arXiv preprint arXiv:2304.02643 (2023)

15. Li, J., Zhang, Z., Zhao, H.: Self-prompting large language models for open-domain qa. arXiv preprint arXiv:2212.08635 (2022)

16. Ma, J., Wang, B.: Segment anything in medical images. arXiv preprint arXiv:2304.12306 (2023)

17. Makarevich, A., Farshad, A., Belagiannis, V., Navab, N.: Metamedseg: volumetric meta-learning for few-shot organ segmentation. arXiv preprint arXiv:2109.09734 (2021)

18. Mattjie, C., .: Exploring the zero-shot capabilities of the segment anything model (sam) in 2d medical imaging: a comprehensive evaluation and practical guideline. arXiv preprint arXiv:2305.00109 (2023)

19. Mohapatra, S., Gosai, A., Schlaug, G.: Brain extraction comparing segment anything model (sam) and fsl brain extraction tool. arXiv preprint arXiv:2304.04738 (2023)

20. OpenAI: Gpt-4 technical report (2023)

21. Pedregosa, F., et al.: Scikit-learn: machine learning in Python. J. Mach. Learn. Res. **12**, 2825–2830 (2011)

22. Ramesh, A., et al.: Zero-shot text-to-image generation (2021)

23. Ronneberger, O., Fischer, P., Brox, T.: U-Net: convolutional networks for biomedical image segmentation. In: Navab, N., Hornegger, J., Wells, W.M., Frangi, A.F. (eds.) MICCAI 2015. LNCS, vol. 9351, pp. 234–241. Springer, Cham (2015). https://doi.org/10.1007/978-3-319-24574-4_28

24. Singh, R., Bharti, V., Purohit, V., Kumar, A., Singh, A.K., Singh, S.K.: Metamed: few-shot medical image classification using gradient-based meta-learning. Pattern Recogn. **120**, 108111 (2021)

25. Sun, L., et al.: Few-shot medical image segmentation using a global correlation network with discriminative embedding. Comput. Biol. Med. **140**, 105067 (2022)

26. Tschandl, P., Rosendahl, C., Kittler, H.: Data descriptor: the ham10000 dataset, a large collection of multi-source dermatoscopic images of common pigmented skin lesions. Sci. Data **5**(1) (2018)

27. Wang, R., Zhou, Q., Zheng, G.: Few-shot medical image segmentation regularized with self-reference and contrastive learning. In: International Conference on Medical Image Computing and Computer-Assisted Intervention, pp. 514–523. Springer (2022). https://doi.org/10.1007/978-3-031-16440-8_49

28. Wu, J., et alT.: Medical sam adapter: adapting segment anything model for medical image segmentation. arXiv preprint arXiv:2304.12620 (2023)

29. Zhang, K., Liu, D.: Customized segment anything model for medical image segmentation. arXiv preprint arXiv:2304.13785 (2023)

30. Zhou, T., Zhang, Y., Zhou, Y., Wu, Y., Gong, C.: Can sam segment polyps? arXiv preprint arXiv:2304.07583 (2023)

Author Index

L. Koch et al. (Eds.): DART 2023, LNCS 14293, pp. 169–170, 2024.
https://doi.org/10.1007/978-3-031-45857-6

Printed in the United States
by Baker & Taylor Publisher Services